高等学校建筑类专业英语规划教材

建筑环境与设备工程专业

(建筑环境与能源应用工程专业)

李安桂　主　编
闫秋会　南晓红　张孙孝　副主编

中国建筑工业出版社

图书在版编目（CIP）数据

建筑环境与设备工程专业/李安桂主编.—北京：中国建筑工业出版社，2011.9（2022.6重印）
高等学校建筑类专业英语规划教材.建筑环境与能源应用工程专业
ISBN 978-7-112-13411-3

Ⅰ.①建… Ⅱ.①李… Ⅲ.①建筑工程—环境管理—高等学校—教材②房屋建筑设备—高等学校—教材 Ⅳ.①TU—023②TU8

中国版本图书馆 CIP 数据核字（2011）第 151545 号

责任编辑：张文胜　田启铭
责任设计：李志立
责任校对：陈晶晶　王雪竹

高等学校建筑类专业英语规划教材
建筑环境与设备工程专业
（建筑环境与能源应用工程专业）

李安桂　主　编
闫秋会　南晓红　张孙孝　副主编

*

中国建筑工业出版社出版、发行（北京西郊百万庄）
各地新华书店、建筑书店经销
华鲁印联（北京）科贸有限公司制版
北京建筑工业印刷厂印刷

*

开本：787×1092 毫米　1/16　印张：13¾　字数：343 千字
2011 年 8 月第一版　2022 年 6 月第八次印刷
定价：35.00 元
ISBN 978-7-112-13411-3
（21138）

版权所有　翻印必究
如有印装质量问题，可寄本社退换
（邮政编码　100037）

前　言

　　本书旨在培养建筑环境与设备工程（建筑环境与能源利用工程）专业学生的科技英语阅读、翻译和写作能力，提高学生以英语为工具获取专业所需信息的能力，为学生日后的工作、科学研究以及国际学术交流等打下良好的英语基础。

　　本书内容体系、课文选材考虑了学术性、实用性、可读性等方面。期望本书作为敲门之石，帮助大学生建立工程师素质、专业能力、未来进一步深造所关心的国外高校研究方向动态的成功之桥，学而有所思、有所得、有所悟。本书主要内容直接选自英、美等国家的高等教育年鉴、原版教科书和相关大学的课程体系设置，涉及工程热力学、传热学、流体力学、供热、空调、制冷、热泵、通风、锅炉、可再生能源等相关知识。结合作者多年的教学经验，在保持以往同类教材优点的基础上，本书在内容选取、体系设置等方面力争具有以下特色：

　　（1）内容选取上，既注重专业基础内容，又追踪暖通空调的发展历史，并关注目前的研究热点，力争实现"启发性"、"知识性"和"前瞻性"的有机结合；

　　（2）内容安排上，每个单元后均配有生词、短语、注释内容以及练习题，特别注重学生写作能力的培养；

　　（3）本书适当增加了图片信息，加强了直观效果，以期激发读者的阅读兴趣；

　　（4）本书附录包括了本专业涉及的国际期刊、学术组织、国际会议、专业术语、相关规范的翻译以及国际著名大学的课程设置等内容，对大学生日后的工作和进一步深造有一定的参考价值。

　　本书力图实现"启发性"、"可读性"、"前瞻性"以及"书目有留存价值"的有机融合，旨在使学生专业英语的阅读能力、翻译能力、写作能力及科学思维能力能够迈上新的台阶。

　　本书编写分工如下：李安桂（西安建筑科技大学）任主编，负责教材体系、选材衔接及内容科学设置等工作，并编写 Lesson 1 以及附录 A～附录 D。闫秋会（西安建筑科技大学）编写了 Lesson2～Lesson6 以及 Lesson17，南晓红（西安建筑科技大学）编写了 Lesson7～Lesson11 以及 Lesson16，张孙孝（长安大学）编写了 Lesson12～Lesson15。全书由李安桂教授统稿。

　　本书在编写过程中，研究生刘菊、王罡、王伟、甄亚曼、宋艳艳和刘永娟等作了资料性协助工作，在此一并表示感谢。

　　本书得到"西安建筑科技大学专业外语系列重点教材"教改项目的资助。本书可作为建筑环境与设备工程专业及相关专业本科生专业英语教材，同时也可以作为相关专业教师、工程技术人员和科研人员的参考书。由于编者水平有限，编写中难免有不妥之处，恳请广大读者对本教材的缺点错误予以斧正。

CONTENTS

Part 1 Reading Courses of Occupational English for HVAC（专业英语教程） ············ 1

  Lesson 1 Requirements for an Engineer and Brief Introduction to Engineering
          （工程师职业素养） ··· 2

  Lesson 2 Engineering Thermodynamics（工程热力学） ··································· 11
    I. Text ··· 11
    II. Words and Expressions ·· 16
    III. Notations ·· 17
    IV. Exercises ··· 18

  Lesson 3 Heat Transfer（传热学） ··· 20
    I. Text ··· 20
    II. Words and Expressions ·· 25
    III. Notations ·· 26
    IV. Exercises ··· 27

  Lesson 4 Fluid Mechanics（流体力学） ··· 29
    I. Text ··· 29
    II. Words and Expressions ·· 35
    III. Notations ·· 35
    IV. Exercises ··· 36
    Extensive Reading ··· 37
    History of Thermodynamics ·· 37
    A Brief History of Computational Fluid Dynamics (CFD) ····················· 39

  Lesson 5 Heating Engineering（供暖工程） ·· 43
    I. Text ··· 43
    II. Words and Expressions ·· 49
    III. Notations ·· 50
    IV. Exercises ··· 51

  Lesson 6 Gas Supply（燃气供应） ··· 53
    I. Text ··· 53
    II. Words and Expressions ·· 59
    III. Notations ·· 59
    IV. Exercises ··· 61
    Extensive Reading ··· 61
    The Invention of District Heating ··· 61

A Brief History of Radiant Panel Heating ·· 62
Lesson 7　Ventilation System（通风系统）··· 65
　　　Ⅰ. Text ·· 65
　　　Ⅱ. Words and Expressions ··· 71
　　　Ⅲ. Notations ··· 72
　　　Ⅳ. Exercises ··· 72
Lesson 8　Air-Conditioning System（空调系统）·· 75
　　　Ⅰ. Text ·· 75
　　　Ⅱ. Words and Expressions ··· 81
　　　Ⅲ. Notations ··· 82
　　　Ⅳ. Exercises ··· 83
Lesson 9　Boiler System（锅炉系统）··· 85
　　　Ⅰ. Text ·· 85
　　　Ⅱ. Words and Expressions ··· 90
　　　Ⅲ. Notations ··· 91
　　　Ⅳ. Exercises ··· 93
Lesson 10　Refrigeration Systems（制冷系统）·· 95
　　　Ⅰ. Text ·· 95
　　　Ⅱ. Words and Expressions ··· 100
　　　Ⅲ. Notations ··· 101
　　　Ⅳ. Exercises ··· 101
Lesson 11　Description of Ground-Source Types for Heat Pump
　　　　　（地源热泵）·· 103
　　　Ⅰ. Text ·· 103
　　　Ⅱ. Words and Expressions ··· 109
　　　Ⅲ. Notations ··· 109
　　　Ⅳ. Exercises ··· 110
　　　Reading Material ··· 111
　　　A Brief History of Air Conditioning ··· 111
　　　A Historical Perspective of Ventilation ·· 113
　　　A Modern Perspective of Ventilation ··· 114
Lesson 12　Building Automation Systems（建筑自动化系统）····················· 116
　　　Ⅰ. Text ·· 116
　　　Ⅱ. Words and Expressions ··· 122
　　　Ⅲ. Notations ··· 124
　　　Ⅳ. Exercises ··· 125
Lesson 13　Particle Image Velocimetry（粒子图像测速）······························ 126
　　　Ⅰ. Text ·· 126
　　　Ⅱ. Words and Expressions ··· 132

  III. Notations ······ 133
  IV. Exercises ······ 134
 Lesson 14 Solar Air Conditioning（太阳能空调）······ 136
  I. Text ······ 136
  II. Words and Expressions ······ 140
  III. Notations ······ 142
  IV. Exercises ······ 143
 Lesson 15 Introduction to Thermal Comfort（热舒适性简介）······ 145
  I. Text ······ 145
  II. Words and Expressions ······ 151
  III. Notations ······ 152
  IV. Exercises ······ 153
  Extensive Reading ······ 155
  Top 10 Ways Homeowners Can Ensure Good IAQ ······ 155
  A Brief History of Particle Image Velocimetry ······ 155

Part 2 Ability Enhancement of Occupational English for HVAC
  （专业英语应用能力拓展）······ 157
 Lesson 16 English Writing Guideline（英文科研论文写作简介）······ 158
 Lesson 17 Information of International Organization, Journals and Conferences
  （著名国际组织、国际期刊、国际会议）······ 172

Part 3 Appendix（附录）······ 178
 Appendix A Basic Terminology for HVAC（专业术语荟萃）······ 179
 Appendix B Introduction to SCI、EI and ISTP（SCI、EI 与 ISTP 检索
  工具简介）······ 195
 Appendix C Main Courses for HVAC in Some Famous Universities and
  Institutes（相关大学 HVAC 课程一览表）······ 197
 Appendix D Key Chinese Laws, Regulations and Standards Pertaining to
  Urban Housing Development in Xi'an（HVAC 相关中国法律、
  规范及标准）······ 208
 参考文献 ······ 213

Part 1 Reading Courses of Occupational English for HVAC
(专业英语教程)

Lesson 1 Requirements for an Engineer and Brief Introduction to Engineering（工程师职业素养）

What is Engineering?

[1] Engineering is the art of applying scientific and mathematical principles, experience, judgment, and common sense to make things that benefit people. Engineers design bridges and important medical equipment as well as processes for cleaning up toxic spills and systems for mass transit. In other words, engineering is the process of producing a technical product or system to meet a specific need.

[2] Engineers have many different types of jobs to choose from, including research, design, analysis, development, testing, and sales positions. If you are interested in discovering new knowledge, you might consider a career as a research engineer. If you are imaginative and creative, design engineering may be for you. The work of analytical engineers most closely resembles what you do in your mathematics and science classes. If you like laboratory courses and conducting experiments, look into becoming a development engineer. Sales engineering could be a good choice if you are persuasive and like working with people.

[3] Engineering work is also organized by traditional academic fields of study. The five largest of these are chemical, civil, electrical, industrial, and mechanical engineering. There are also more specialized engineering fields, including aerospace, ocean, nuclear, biomedical, and environmental engineering.

What is HVAC?

[4] You've probably heard of the term from different contractors, engineers, or perhaps colleagues and business partners; but you're still wondering what the initialism means. Well, HVAC ("H-V-A-C" or "H-VAC") stands for Heating, Ventilation, and Air-Conditioning—three closely related fundamental functions found in homes, offices, and other building structures.

[5] The beginning of HVAC is not clear, though as early as second century, a lot of Roman cities were using a central heating system known as hypocaust. This is further popularized during the Industrial Revolution as big factories used it. Now most modern buildings that you see have integrated HVAC.

[6] The HVAC system is also known as climate control. This is because these three functions are essential in maintaining comfort in every dwelling.

[7] The primary use of HVAC is to regulate room temperature, humidity, and air flow, ensuring that such elements remain within their acceptable ranges. Effective control of such factors minimizes health-related risks. A very humid atmosphere impairs the body's ability to regulate body temperature as it prevents the evaporation of sweat. High humidity also decreases physical strength, which usually leads to fatigue. An unhealthy surrounding

can also affect people's thinking abilities. Hypothermia, heat stroke, and hyperpyrexia, among others, are some of the illnesses that may also occur.

Three Functions of HVAC

[8] Heating is significant in maintaining adequate room temperature especially during colder weather conditions. There are two classifications of heating: local and central. The latter is more commonly used because it is more economical. Furnace or boiler, heat pump, and radiator make up the heating system.

[9] Ventilation, on the other hand, is associated with air movement. There are many types of ventilation, but they all function similarly. Ventilation is necessary to allow carbon dioxide to go out and oxygen to get in, making sure that people are inhaling fresh air. Stagnant air causes the spreading of sickness, usually airborne, and allergies. But it is also essential to maintain an efficient ventilation system, especially in the attics. Insufficient ventilation usually promotes the growth of bacteria and fungi such as molds because of high humidity. It will also decrease the effectiveness of rafter and roof sheathing insulation because of water vapor condensation.

[10] The air-conditioning system controls the heat as well as ventilation. They often come in different sizes. Most air conditioners have large air ducts, so it is better to check out the building first to see if they can be installed. Or else, you can use the split system or remote coils. It is necessary, though, that air ducts are properly cleaned. Pathogens thrive in dirty air ducts. Return-air grills are also vulnerable to chemical, microbiological, and radiological elements. Thus, HVAC return-air grill height should be that it is not accessible but visible for any observation.

The Future of HVAC

[11] How has technology changed in the HVAC field? Well, using PLCs (Programmable Logic Controllers) in HVAC is the trend nowadays. But a great deal of development of the HVAC system lies on the ever-changing technology and continuous innovation. Companies are adopting wireless technology after they found out that networking HVAC controllers, which often use sensors, can eventually cut installation and labor costs. A lot of engineers are also focused on further improving this technology through the use of mesh wireless setup, which will work for both the wireless sensor and wireless controller networks. The only downside of this could probably be the risk of being exposed to RF (Radio Frequency) radiation.

[12] The installation of an HVAC system is imperative if we want to achieve maximum comfort and be healthy in our homes, office spaces, or other building facilities. But you also need to consider the building size in installing an HVAC system. Optimum efficiency and comfort level are best achieved if the system is appropriate for the size. After all, any ineffective system usually means more incurred costs in the future. You should also see to it that HVAC is carefully integrated to the overall building design so other aspects needed for proper operations, such as cabling, are not sacrificed.

What is Civil Engineering and HVAC Engineering?

[13] Civil engineering is a remarkably broad field of study serving people by designing, constructing and maintaining the infrastructure of society including buildings, bridges, highways, airports, and harbors. Civil engineers impact the quality of the built environment and the quality of our land, water, and air resources. Students can elect to prepare for professional careers such as structural, environmental, transportation, construction, hydraulic, geotechnical, or surveying engineers. The curriculum accommodates this breadth by providing a fundamental set of required courses complemented by sufficient flexibility to allow students to concentrate portions of their studies on the special areas that are of particular interest to them.

Ten great reasons why you will love it!

What is an engineer?

[14] Engineers use their imagination and analytical skills to invent, design, and build things that matter. They are team players with independent minds who ask, "How can we develop a better recycling system to protect the environment, design a school that can withstand an earthquake, or create cutting-edge special effects for the movies?" By dreaming up creative and practical solutions, engineers are changing the world all the time.

Nature of the Engineer Work

[15] Engineers apply the principles of science and mathematics to develop economical solutions to technical problems. Their work is the link between scientific discoveries and the commercial applications that meet societal and consumer needs.

[16] Many engineers develop new products. During this process, they consider several factors. For example, in developing an industrial robot, engineers precisely specify the functional requirements; design and test the robot's components; integrate the components to produce the final design; and evaluate the design's overall effectiveness, cost, reliability, and safety. This process applies to the development of many different products, such as chemicals, computers, power plants, helicopters, and toys.

[17] In addition to design and development, many engineers work in testing, production, or maintenance. These engineers supervise production in factories, determine the causes of component failure, and test manufactured products to maintain quality. They also estimate the time and cost to complete projects. Supervisory engineers are responsible for major components or entire projects.

[18] Engineers use computers extensively to produce and analyze designs; to simulate and test how a machine, structure, or system operates; to generate specifications for parts; and to monitor product quality and control process efficiency. Nanotechnology, which involves the creation of high-performance materials and components by integrating atoms and molecules, also is introducing entirely new principles to the design process.

[19] Most engineers specialize. Following are details on the 17 engineering specialties covered in the Federal Government's Standard Occupational Classification (SOC) system.

Numerous other specialties are recognized by professional societies, and each of the major branches of engineering has numerous subdivisions. Civil engineering, for example, includes structural and transportation engineering, and materials engineering includes ceramic, metallurgical, and polymer engineering. Engineers also may specialize in one industry, such as motor vehicles, or in one type of technology, such as turbines or semiconductor materials.

Work Environment.

[20] Most engineers work in office buildings, laboratories, or industrial plants. Others may spend time outdoors at construction sites and oil and gas exploration and production sites, where they monitor or direct operations or solve onsite problems. Some engineers travel extensively to plants or worksites here and abroad.

[21] Many engineers work a standard 40-hour week. At times, deadlines or design standards may bring extra pressure to a job, requiring engineers to work longer hours.

- **Love your work, and live your life too!**

[22] Engineer is an exciting, rewarding career choice for young men and women. Engineering is an exciting profession, but one of its greatest advantages is that it will leave you time for all the other things in your life that you love!

- **Be creative**

[23] Engineering is a great outlet for the imagination—the perfect field for independent thinkers.

- **Work with great people**

[24] Engineering takes teamwork, and you'll work with all kinds of people inside and outside the field. Whether they're designers or architects, doctors or entrepreneurs, you'll be surrounded by smart, inspiring people.

- **Solve problems, design things that matter**

[25] Come up with solutions no one else has thought of. Make your mark on the world.

- **Never be bored**

[26] Creative problem solving will take you into uncharted territory, and the ideas of your colleagues will expose you to different ways of thinking. Be prepared to be fascinated and to have your talents stretched in ways you never expected.

- **Make a big salary**

[27] Engineers not only earn lots of respect, but they're highly paid. Even the starting salary for an entry-level job is impressive!

- **Enjoy job flexibility**

[28] An engineering degree offers you lots of freedom in finding your dream job. It can be a launching pad for jobs in business, design, medicine, law, and government. To employers or graduate schools, an engineering degree reflects a well-educated individual who has been taught ways of analyzing and solving problems that can lead to success in all

kinds of fields.

- **Travel**

[29] Field work is a big part of engineering. You may end up designing a skyscraper in London or developing safe drinking-water systems in Asia. Or you may stay closer to home, working with a nearby high-tech company or a hospital.

- **Make a difference**

[30] Everywhere you look you'll see examples of engineering having a positive effect on everyday life. Cars are safer, sound systems deliver better acoustics, medical tests are more accurate, and computers and cell phones are a lot more fun! You'll be giving back to your community.

- **Change the world**

[31] Imagine what life would be like without pollution controls to preserve the environment, life-saving medical equipment, or low-cost building materials for fighting global poverty. All this takes engineering. In very real and concrete ways, engineers save lives, prevent disease, reduce poverty, and protect our planet.

[32] Today, just 20 percent of undergraduate engineering students are women. Even more astounding is the number of women engineers in the professional workforce - less than ten percent! Engineer Your Life is an unprecedented awareness and outreach program designed to encourage young women to choose engineering as a career and to develop a new generation of role models for those already in the field.

What Engineers Do: A Look Behind the Scenes

[33] Typical Progression:

- Orientation and some reading
- Small assigned task with supervision
- Small assigned task with minimum supervision
- Larger assigned task with supervision
- Large assigned task with technician support
- Help choose your own task: Define your own support, schedule
- Define a task and work with others to get it done
- Define a significant task and lead a group of peers
- Become a local expert in your area of expertise
- Interact with other groups and secure their cooperation to accomplish a task
- Help define a major piece of work: Lay out the plan and execute it
- Publish/invent/document significant ideas
- Become the major engineer on a product or major part of a product
- Choose a management career: Manage a department, or Secure support for your own project
- Choose a technical career: Become a recognized expert at your site and beyond, or Consult widely in your area of expertise

- Lead major task forces
- Move up the corporate ladder

[34] Solve Design Problems:
Fuzzy definition of task ("Figure out what needs doing and do it!")
- Factor in time constraint ("If I wanted it tomorrow, I'd ask you tomorrow!")
- Factor in cost constraint (Cheaper than anyone ever did it before)
- Factor in constrained resources (few people, equipment not available)
- Balance priority with other duties ("Don't let your other work slip!")
- Learn where to seek help: Develop a support group.

[35] Schedules:
- State resource assumptions (people, equipment)
- Do two schedules: Working back from end date, Working forward from start date, Seek a reasonable balance
- State dependencies and exposures
- Draw up schedule with tasks, start/end dates
- Review schedule with peers for sanity
- Submit to manager for review
- Rework schedule (Don't commit to insanity!)
- Final review with manager
- Do it!: Track your schedule, Make notes of what you forgot for next time, Get good at scheduling (Your boss will love you!)

[36] Product Planning:
- Market requirements (Who wants it, what functions are important?)
- Competitive analysis (Who is competition, what performance and price?)
- When could the product be announced?
- What countries? Language translations.
- Development plan: Assign design team, Budget/resource estimate, Functional specifications, Major check points, Development schedule (components/prototypes), Test plans with schedule, Test model build/debug, Test schedule, Documentation schedule, Problem resolution and redesign time, Manufacturing plan, Marketing plan, Human factors, Costs/cost reduction plan
- Reliability/service cost estimate: Monthly usage estimate, Repair actions per month, Duration of repair action, Service aids
- Market plan: Who will sell it, Who will buy it, How many will be sold, Advertising, Sales incentives, Order/delivery method, Service/maintenance plan, Marketing/development commitment to plan, Check points, Marketing literature, Early demo hardware
- Manufacturing plan: Where to be manufactured, How many (by month), Manufacturing cost estimate, Packaging and ship group, Distribution, Commitment agreements, Budget (by quarter/year), Capital expenses/schedule (equipment), Head count buildup

(and skills)
- Test schedule and early model quantities/assignments
- Announce date and general availability dates
- Reviews by everyone involved (non-congruency and resolution)
- Continual negotiation
- Profit/revenue plan
- Safety/security/EO plans
- Budget/space/resource/schedule/equipment plans: Justify everything, Multiple iterations, Fight for your share
- Personnel appraisals/promotions/awards/reprimands/dismissals
- Presentations to upper management on every subject
- Keep employees appropriately informed
- Maintain troop morale throughout tough times
- Look calm and reassure the troops when the word is caving in
- Plan ahead constantly for contingencies (and insanity)
- Protect troops/turf against raiders and upper management gyrations
- Work within budget/manpower/schedule/space constraints (ever changing)
- Get others to do work for you, preferably out of their budget
- Write nice notes to managers of folks who help you: They will love to help you again, Your note help justify raises/promotions
- Calmly take the blame for everything that happens: By definition you are responsible, Excuses do not impress anyone
- Know everything about anything ("Ask your manager!")
- Great responsibility with surprisingly little authority: Succeed in spite of it all, Find ways around the system (its your job), Ask forgiveness (not permission), Be unbelievably creative, Don't whine
- Praise your good people to the sky! Get them incredible raises and promotions!

[37] Senior Technical Tasks:
- Budgeting: People, Money, Capital equipment, Space, Salaries, Support from other groups
- Set a professional tone for the department
- Encourage colleagues
- Mentor younger colleagues
- Publish papers, formal and informal
- Teach courses
- Patent applications
- Set technical direction for the department, site
- Lead Technical Presentations: Department, Management, Tech Society, Executives
- Lead design reviews

- Lead/partcipate in program audits
- Lead Task Forces (Serious technical problems): Listen and absorb info quickly, Recommend actions, Create and defend action plan, Get resources, Assign tasks, Monitor results daily, Daily/Weekly Status Reports, Management/Executive Presentations, Hold off the Doomsayers, Final recommendations (report)
- Consult widely (formal, phone calls, meetings)
- Participate in advanced technical planning
- Continue to study and grow in your field of expertise
- Broaden your areas of expertise-tackle new things

Engineering Specialty Employment in Key Industries

[38] About 37 percent of engineering jobs were found in manufacturing industries and another 28 percent were in the professional, scientific, and technical services sector, primarily in architectural, engineering, and related services. Many engineers also worked in the construction, telecommunications, and wholesale trade industries.

[39] Federal, State, and local governments employed about 12 percent of engineers in 2006. About half of these were in the Federal Government, mainly in the U. S. Departments of Defense, Transportation, Agriculture, Interior, and Energy, and in the National Aeronautics and Space Administration. Most engineers in State and local government agencies worked in highway and public works departments. In 2006, about 3 percent of engineers were self-employed, many as consultants.

[40] Engineers are employed in every State, in small and large cities and in rural areas. Some branches of engineering are concentrated in particular industries and geographic areas—for example, petroleum engineering jobs tend to be located in areas with sizable petroleum deposits, such as Texas, Louisiana, Oklahoma, Alaska, and California. Others, such as civil engineering, are widely dispersed, and engineers in these fields often move from place to place to work on different projects.

[41] Engineers are employed in every major industry. The industries employing the most engineers in each specialty are given in table 1, along with the percent of occupational employment in the industry.

Percent concentration of engineering specialty employment in key industries, 2006　Table 1

Specialty	Industry	Percent
Aerospace engineers	Aerospace product and parts manufacturing	49
Agricultural engineers	Food manufacturing	25
	Architectural, engineering, and related services	15
Biomedcal engineers	Medical equipment and supplies manufacturing	20
	Scientific research and development services	20
Chemical engineers	Chemical manufacturing	29

continued

Specialty	Industry	Percent
	Architectural, engineering, and related services	15
Civil engineers	Architectural, engineering, and related services	49
Computer hardware engineers	Computer and electronic product manufacturing	41
	Computer systems design and related services	19
Electrical engineers	Architectural, engineering, and related services	21
Electronics engineers, except computer	Computer and electronic product manufacturing	26
	Telecommunications	15
Environmental engineers	Architectural, engineering, and related services	29
	State and local government	21
Health and safety engineers, except mining safety engineers and inspectors	State and local government	10
Industrial engineers	Transportation equipment manufacturing	18
	Machinery manufacturing	8
Marine engineers and naval architects	Architectural, engineering, and related services	29
Materials engineers	Primary metal manufacturing	11
	Semiconductor and other electronic component manufacturing	9
Mechanical engineers	Architectural, engineering, and related services	22
	Transportation equipment manufacturing	14
Mining and geological engineers, including mining safely engineers	Mining	58
Nuclear engineers	Research and development in the physical, engineering, and life sciences	30
	Electric power generation, transmission and distribution	27
Petroleum engineers	Oil and gas extraction	43

Lesson 2 Engineering Thermodynamics (工程热力学)

I. Text

[1] Thermodynamics can be defined as the science of energy. It is broadly interpreted to include all aspects of energy and energy transformations, including power generation, refrigeration, and relationship among the properties of matter. This chapter covers the first and second laws of thermodynamics and presents methods for calculating thermodynamic properties. The first part introduces some basic concepts of thermodynamics. The second part reviews the first and second laws of thermodynamics.

Basic Concepts of Thermodynamics

[2] A thermodynamic system, or simply a system, is defined as a quantity of matter or a region in space chosen for study. The mass or region outside the system is called the surroundings. The real or imaginary surface that separates the system from its surroundings is called the boundary. The boundary of a system can be fixed or movable.

[3] Two master concepts operate in any thermodynamic system, energy and entropy. **Entropy measures the molecular disorder of a system. The more shuffled a system is, the greater its entropy; conversely, an orderly or unmixed configuration is one of low entropy.**

Energy

[4] Energy is the capacity for producing an effect, and can exist in numerous forms such as thermal, mechanical, kinetic, potential, electric, magnetic, chemical, and nuclear, and their sum constitutes the total energy E of a system.

Thermal (internal) energy, u—the energy possessed by a system caused by the motion of molecules and/or intermolecular forces.

Potential energy (E_p) —the energy possessed by a system caused by the attractive forces existing between molecules, or the elevation of the system.

$$E_p = mgz \qquad (1)$$

where g is the gravitational acceleration and z is elevation of the center of gravity of a system relative to some arbitrarily selected reference plane.

Kinetic energy (E_k) —the energy possessed by the velocity of molecules.

$$E_k = mv^2/2 \qquad (2)$$

where v is the velocity of a fluid stream crossing the system boundary.

Chemical energy (E_c) —energy possessed by the arrangement of atoms composing the molecules.

Nuclear (atomic) energy, an energy possessed by the system from the cohesive forces holding protons and neutrons together as the atom's nucleus.

[5] Energy can be transferred to or from a closed system (a fixed mass) in two distinct

forms: heat and work. **Heat is defined as the form of energy that is transferred between two systems (or a system and its surrounding) by virtue of temperature difference. That is, an energy interaction is heat only if it takes place because of a temperature difference.**

[6] Work is also a form of energy transferred like heat and, therefore, has energy units such as kJ. Work is the energy transfer associated with force acting through a distance. It transfers energy across the boundaries of systems with differing pressures (or force of any kind).

[7] There are several different ways of doing work, each in some way related to a force acting through a distance. In elementary mechanics, the work done by a constant force F on a body displaced a distance s in the direction of the force is given by

$$W = Fs \text{ (kJ)} \qquad (3)$$

If the force F is not constant, the work done is obtained by adding the differential amounts of work,

$$W = \int_1^2 F \mathrm{d}s \text{ (kJ)} \qquad (4)$$

[8] Flow work is energy carried into or transmitted across the system boundary because a pumping process occurs somewhere outside the system, causing fluid to enter the system. It can be more easily understood as the work done by the fluid just outside the system on the adjacent fluid entering the system to force or push it into the system. Flow work also occurs as fluid leaves the system.

$$\text{Flow work (per unit mass)} = pV \qquad (5)$$

where p is the pressure and V is the specific volume, or the volume displaced per unit mass evaluated at the inlet or exit.

[9] Mechanical or shaft work W is the energy delivered or absorbed by a mechanism, such as a turbine, air compressor, or internal combustion engine.

Entropy

[10] Clausius realized in 1865 that he had discovered a new thermodynamic property, and he chooses to name this property entropy. It is designated S and is defined as

$$\mathrm{d}S = \left(\frac{\delta Q}{T}\right)_{\text{int rev}} \text{ (kJ/K)} \qquad (6)$$

The entropy change of a system during a process can be determined by integrating Eqnation (6) between the initial and the final states:

$$\Delta S = S_2 - S_1 = \int_1^2 \left(\frac{\delta Q}{T}\right)_{\text{int rev}} \text{ (kJ/K)} \qquad (7)$$

The entropy generated during a process is called entropy generation and is denoted by S_{gen}. Noting that the difference between the entropy change of a closed system and the entropy transfer is equal to entropy generation:

$$\Delta S_{\text{sys}} = S_2 - S_1 = \int_1^2 \frac{\delta Q}{T} + S_{\text{gen}} \qquad (8)$$

[11] Note that the entropy generation S_{gen} is always a positive quantity or zero. Its

value depends on the process, and thus it is not a property of the system. Also, in the absence of any entropy transfer, the entropy change of a system is equal to the entropy generation. For an isolated system (or simply an adiabatic closed system), the heat transfer is zero,

$$\Delta S_{\text{isolated}} \geqslant 0 \tag{9}$$

[12] **This equation can be expressed as the entropy of an isolated system during a process always increases or, in the limiting case of a reversible process, remains constant. In other words, it never decreases. This is known as the increase of entropy principle.** Note that in the absence of any heat transfer, entropy change is due to irreversibility only, and their effect is always to increase entropy.

[13] The increase of entropy principle does not imply that the entropy of a system cannot decrease. The entropy change of a system can be negative during a process, but entropy generation cannot. The increase of entropy principle can be summarized as follows:

$$S_{\text{gen}} = \begin{cases} >0 & \text{irreversible process} \\ =0 & \text{reversible process} \\ <0 & \text{impossible process} \end{cases} \tag{10}$$

This relation serves as a criterion in determining whether a process is reversible, irreversible, or impossible.

Properties of a System

[14] Any characteristic of a system is called a property. Some familiar properties are pressure P, temperature T, volume V, and mass m. The list can be extended to include less familiar ones such as viscosity, thermal conductivity, modulus of elasticity, thermal expansion coefficient, electric resistivity, and even velocity and elevation.

Process and Cycles

[15] Any change that a system undergoes from one equilibrium state to another is called a process, and the series of states through which a system passes during a process is called the path of the process. To describe a process completely, one should specify the initial and final states of the process, as well as the path it follows, and the interactions with the surroundings. A system is said to have undergone a cycle if it returns to its initial state at the end of the process. That is, for a cycle the initial and final states are identical.

First Law of Thermodynamics

[16] The first law of thermodynamics is simply a statement of the conservation of energy principle, and it asserts that total energy is a thermodynamic property. The first law of thermodynamics, also known as the conservation of energy principle, provides a sound basis for studying the relationships among the various forms of energy and energy interactions. The first law of thermodynamics states that energy can be neither created nor destroyed; it can only change forms. Therefore, every bit of energy should be accounted for during a process. A system is said to have undergone a cycle if it returns to its initial state at the end of the process. That is, for a cycle the initial and final states are identical.

Energy Balance

[17] In the light of the discussions above, the conservation of energy principle may be expressed as follows: **The net change (increase or decrease) in the total energy of the system during a process is equal to the difference between the total energy entering and the total energy leaving the system during that process.**

Mechanisms of Energy Transfer

[18] Noting that energy can be transferred in the forms of heat, work, and mass, and that the net transfer of a quantity is equal to the difference between the amounts transferred in and out, the energy balance can be written more explicitly as

$$E_{in} - E_{out} = (Q_{in} - Q_{out}) + (W_{in} - W_{out}) + (E_{mass,in} - E_{mass,out}) = \Delta E_{system} \tag{11}$$

where the subscripts "in" and "out" denote quantities that enter and leave the system, respectively.

[19] For the general case of multiple mass flows with uniform properties in and out of the system, the energy balance can be written

$$\sum m_{in}\left(u + pV + \frac{v^2}{2} + gz\right)_{in} - \sum m_{out}\left(u + pV + \frac{v^2}{2} + gz\right)_{out} + Q - W$$
$$= \left[m_f\left(u + \frac{v^2}{2} + gz\right)_f - m_i\left(u + \frac{v^2}{2} + gz\right)_i\right]_{system} \tag{12}$$

where subscripts i and f refer to the initial and final states, respectively.

[20] Nearly all important engineering processes are commonly modeled as steady-flow processes. Steady flow signifies that all quantities associated with the system do not vary with time. Consequently,

$$\dot{Q}_{in} + \dot{W}_{in} + \sum \underbrace{\dot{m}_i\left(h_i + \frac{v_i^2}{2} + gz_i\right)}_{\text{for each inlet}} = \dot{Q}_{out} + \dot{W}_{out} + \sum \underbrace{\dot{m}_e\left(h_e + \frac{v_e^2}{2} + gz_e\right)}_{\text{for each exit}} \tag{13}$$

A second common application is the closed stationary system for which the first law equation reduces to

$$Q_{net,in} - W_{net,out} = \Delta E_{system} \tag{14}$$

Second Law of Thermodynamics

[21] The second law of thermodynamics asserts that processes occur in a certain direction (irreversible) and that energy has quality as well as quantity. A process cannot take place unless it satisfies both the first and the second laws of thermodynamics.

The Second Law of Thermodynamics: Kelvin-Planck Statement

[22] A heat engine, even under ideal conditions, must reject some heat to a low-temperature reservoir in order to complete the cycle. That is, no heat engine can convert all the heat it receives to useful work. This limitation on the thermal efficiency of heat engines forms the basis for the Kelvin-Planck statement of the second law of thermodynamics, which is expressed as follows:

It is impossible for any device that operates on a cycle to receive heat from a single reservoir and produce a net amount of work.

[23] That is, a heat engine must exchange heat with a low-temperature sink as well as a high-temperature source to keep operating. **The Kelvin-Planck statement can also be expressed as no heat engine can have a thermal efficiency of 100 percent, or as for a power plant to operate, the working fluid must exchange heat with the environment as well as the furnace.**

[24] Note that the impossibility of having a 100 percent efficient heat engine is not due to friction or other dissipative effects. It is a limitation that applies to both the idealized and the actual heat engines. Later in this chapter, we develop a relation for the maximum thermal efficiency of a heat engine. We also demonstrate that this maximum value depends on the reservoir temperatures only.

The Second Law of Thermodynamics: Clausius Statement

[25] The Clausius statement is related to refrigerators or heat pumps. The Clausius statement is expressed as follows:

It is impossible to construct a device that operates in a cycle and produces no effect other than the transfer of heat from a lower-temperature body to a higher-temperature body.

[26] Both the Kelvin-Planck and the Clausius statements of the second law are negative statements, and a negative statement cannot be proved. Like any other physical law, the second law of thermodynamics is based on experimental observations. To date, no experiment has been conducted that contradicts the second law, and this should be taken as sufficient evidence of its validity.

The Carnot Cycle

[27] **Reversible cycles that consist entirely of reversible process cannot be achieved in practice because the irreversibility associated with each process cannot be eliminated.** However, reversible cycles provide upper limits on the performance of real cycles. Reversible cycles also serve as starting points in the development of actual cycles and are modified as needed to meet certain requirements.

[28] Probably the best known reversible cycle is the Carnot cycle, the theoretical heat engine that operates on the Carnot cycle is called Carnot heat engine. The Carnot cycle is composed of four reversible processes—two isothermal and two adiabatic—and it can be executed either in a closed or a steady-flow system. Consider a closed system that consists of a gas contained in an adiabatic piston-cylinder device. The insulation of the cylinder head is such that it may be removed to bring the cylinder into contact with reservoirs to provide heat transfer. The four reversible processes that make up the Carnot cycle are: reversible isothermal expansion, reversible adiabatic expansion, reversible isothermal compression and reversible adiabatic

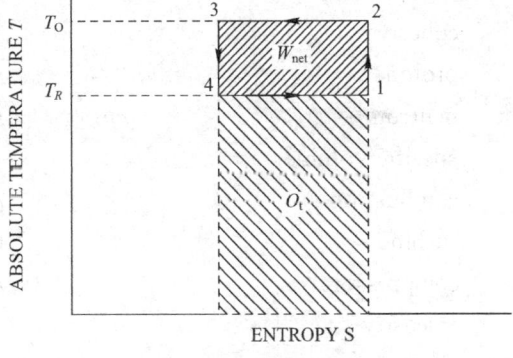

Fig. 1 Carnot Cycle

compression.

[29] Being a reversible cycle, the Carnot cycle is the most efficient cycle operating between two specified temperature limits. **Even though the Carnot cycle cannot be achieved in reality, the efficiency of actual cycles can be improved by attempting to approximate the Carnot cycle more closely.**

[30] The second law of thermodynamics puts limits on the operation of cyclic devices as expressed by the Kelvin-Planck and Clausius statements. A heat engine can not operate by exchanging heat with a single reservoir, and a refrigerator can not operate without a net work input from an external source. Two conclusions pertain to the thermal efficiency of reversible and irreversible heat engines, and they are known as the Carnot principles, expressed as follows:

1. The efficiency of an irreversible heat engine is always less than the efficiency of a reversible one operating between the same two reservoirs.

2. The efficiency of all reversible heat engines operating between the same two reservoirs are the same.

These two statements can be proved by demonstrating that the violation of either statement results in the violation of the second law of thermodynamics.

II. Words and Expressions

thermodynamics	n.	热力学
refrigeration	n.	制冷
entropy	n.	熵（热力学参数）
molecular	a.	分子的
shuffle	v.	搅乱，弄混
thermal	a.	热的
mechanical	a.	机械的
acceleration	n.	（物）加速，加速度
elevation	n.	高度
kinetic	a.	动力（学）的，动力的
cohesive	a.	内聚的
proton	n.	质子
neutron	n.	中子
specific volume		比容
combustion	n.	燃烧
turbine	n.	涡轮机
compressor	n.	压缩机，压气机
viscosity	n.	黏性，黏度
shaft	n.	轴，杆状物
criterion	n.	标准

reversible	a.	可逆的
irreversibility	n.	不可逆性
conductivity	n.	传导性，导电性
modulus of elasticity	n.	弹性模量
furnace	n.	火炉，熔炉
account for	v.	说明，解决
Kelvin-Planck statement		开尔文—普朗克说法
Clausius statement		克劳修斯说法
isothermal	a.	等温的
adiabatic	a.	绝热的
approximate	v.	使…接近，近似
Carnot cycle		卡诺循环
violation	n.	违背，违反

III. Notations

1. Entropy measures the molecular disorder of a system. The more shuffled a system is, the greater its entropy; conversely, an orderly or unmixed configuration is one of low entropy.

熵计量某一给定系统分子的无序程度。系统越紊乱，它的熵就越大。相反，有序或不紊乱的结构是一个低熵系统。

2. Heat is defined as the form of energy that is transferred between two systems (or a system and its surrounding) by virtue of temperature difference. That is, an energy interaction is heat only if it takes place because of a temperature difference.

  by virtue of   由于…

热量是一种能量形式，它是两个系统（或系统与其环境）由于温差而引起传递。也就是说，只是因为温差才产生的能量的相互作用就是热量。

3. This equation can be expressed as the entropy of an isolated system during a process always increases or, in the limiting case of a reversible process, remains constant. In other words, it never decreases. This is known as the increase of entropy principle.

  limiting case   极限情况   be known as   作为…而著名

这个方程可以被表述为孤立系统经过任一过程熵总是增加的，或者在可逆过程这种极限情况下，熵保持不变。换句话说，熵永远不会减少。这就是著名的熵增原理。

4. The net change (increase or decrease) in the total energy of the system during a process is equal to the difference between the total energy entering and the total energy leaving the system during that process.

  be equal to   等于，相等

一个系统经过任一过程总的能量的净变化量（增加或减少）等于这一过程前后进入系统和离开系统的总的能量的差值。

5. The Kelvin-Planck statement can also be expressed as no heat engine can have a

thermal efficiency of 100 percent, or as for a power plant to operate, the working fluid must exchange heat with the environment as well as the furnace.

  as for 就…而言，在…方面

  开尔文说法也可以被描述为，任何热机的热效率都不可能达到100%，就发电厂而言，和锅炉一样，工质必须与环境和锅炉交换热量。

  6. Reversible cycles that consist entirely of reversible process cannot be achieved in practice because the irreversibility associated with each process cannot be eliminated.

  associated with 与…有关系，与…相联系 in practice 事实上，实际上

  可逆循环全部由可逆过程组成，实际上是不可能实现的，因为每一过程的不可逆性是不能消除的。

  7. Even though the Carnot cycle cannot be achieved in reality, the efficiency of actual cycles can be improved by attempting to approximate the Carnot cycle more closely.

  虽然卡诺循环实际上不可能实现，但是实际循环越接近于卡诺循环，其效率越高。

IV. Exercises

  1. Translate the following sentences into Chinese.

  (1) Work is also a form of energy transferred like heat and, therefore, has energy units such as kJ. Work is the energy transfer associated with a force acting through a distance. It transfers energy across the boundaries of systems with differing pressures (or force of any kind).

  (2) The increase of entropy principle does not imply that the entropy of a system cannot decrease. The entropy change of a system can be negative during a process, but entropy generation cannot.

  (3) The first law of thermodynamics, also known as the conservation of energy principle, provides a sound basis for studying the relationships among the various forms of energy and energy interactions. The first law of thermodynamics states that energy can be neither created nor destroyed; it can only change forms.

  (4) Kelvin-Planck statement: It is impossible for any device that operates on a cycle to receive heat from a single reservoir and produce a net amount of work.

  Clausius statement: It is impossible to construct a device that operates in a cycle and produces no effect other than the transfer of heat from a lower-temperature body to a higher-temperature body.

  (5) The Carnot cycle is composed of four reversible processes—two isothermal and two adiabatic—and it can be executed either in a closed or a steady-flow system. Consider a closed system that consists of a gas contained in an adiabatic piston-cylinder device. The insulation of the cylinder head is such that it may be removed to bring the cylinder into contact with reservoirs to provide heat transfer.

  (6) The efficiency of an irreversible heat engine is always less than the efficiency of a reversible one operating between the same two reservoirs. The efficiency of all reversible

heat engines operating between the same two reservoirs are the same.

2. Translate the following sentences into English.

（1）在温差作用下系统与外界传递的能量称为热量，当系统与外界之间达到热平衡时，系统与外界的热量传递随之停止。热量一旦通过界面传入系统，就变成系统储存能的一部分，即内能。显然，热量是与过程特性有关的过程量，而内能是取决于热力状态的状态量。

（2）卡诺循环与卡诺定理在热力学的研究中具有重要的理论和实际意义。它解决了热机热效率的极限值问题，并从原则上提出了提高效率的途径。在相同的热源与冷源之间，卡诺循环的热效率为最高，一切其他实际循环，均低于卡诺循环的热效率。

（3）熵增原理的意义：1）可通过孤立系统的熵增原理判断过程进行的方向；2）熵增原理可作为系统平衡的判据——当孤立系统的熵达到最大值时，系统处于平衡状态；3）熵增原理与过程的不可逆性密切相关，不可逆程度越大，熵增也越大，由此可以定量地评价过程热力学性能的完善性。

Lesson 3　Heat Transfer（传热学）

I. Text

[1] In the simplest of terms, the discipline of heat transfer is concerned with only two things: temperature, and the flow of heat. Temperature represents the amount of thermal energy available, whereas heat flow represents the movement of thermal energy from place to place. Heat transfer is energy transferred because of a temperature difference. Energy moves from a higher-temperature region to a lower-temperature region by one or more of three modes: conduction, radiation, and convection.

Conduction

[2] **Heat transfer by conduction may be though of as the heat transferred through a substance (or combination of substance) from a region of high temperature to a region of low temperature by the progressive exchange of energy between the molecules of the substance.** In the process of transferring heat by conduction, no bodily displacement of the molecules occurs. In the case of metals, however, electron movement greatly assists in heat transfer by conduction.

Fourier Law of Heat Conduction

[3] When a temperature gradient exists within a body, heat energy will flow from the region of high temperature to the region of low temperature. This phenomenon is known as conduction heat transfer, and is described by Fourier's Law (named after the French physicist Joseph Fourier),

$$q = -k \vec{\nabla} T \tag{1}$$

[4] This equation determines the heat flux vector q for a given temperature profile T and thermal conductivity k. The minus sign ensures that heat flows down the temperature gradient.

Steady State 1-Dimensional Heat Conduction

[5] For problems where the temperature variation is only 1-dimensional (say, along the x-coordinate direction), Fourier's Law of heat conduction simples to the scalar equations,

$$q = -k \frac{\partial T}{\partial x} \qquad \dot{Q} = -kA \frac{\partial T}{\partial x} \tag{2}$$

where the heat flux q depends on a given temperature profile T and thermal conductivity k. The minus sign ensures that heat flows down the temperature gradient.

[6] In the above equation on the right, $\dot{Q}$ represents the heat flow through a defined cross-sectional area A, measured in watts,

$$\dot{Q} = \int_A q \cdot dA \tag{3}$$

Integrating the 1D heat flow equation through a material's thickness Dx gives,

$$\dot{Q} = \frac{kA}{\Delta x}(T_1 - T_2) \tag{4}$$

where T_1 and T_2 are the temperatures at the two boundaries.

Convection

[7] Heat energy transfers between a solid and a fluid when there is a temperature difference between the fluid and the solid. This is known as "convection heat transfer". Generally, convection heat transfer can not be ignored when there is a significant fluid motion around the solid.

Natural Convection

[8] Natural convection occurs when a system becomes unstable and therefore begins to mix by the movement of mass. **A common observation of convection is of thermal convection in a pot of boiling water, in which the hot and less-dense water on the bottom layer moves upwards, and the cool and more dense water near the top of the pot likewise sinks.** The onset of natural convection is determined by the Rayleigh number (Ra).

[9] Natural convection will be more likely and more rapid with a greater variation in density between the two fluids, a larger acceleration due to gravity that drives the convection, and a larger distance through the convecting medium. Convection will be less likely and less rapid with more rapid diffusion (thereby diffusing away the gradient that is causing the convection) and a more viscous fluid.

Forced Convection

[10] Forced convection is a type of heat transport in which fluid motion is generated by an external source (like a pump, fan, suction device, etc.). Forced convection is often encountered by engineers designing or analyzing heat exchangers, pipe flow, and flow over a plate at a different temperature than the stream. However, in any forced convection situation, some amount of natural convection is always present whenever there are g-forces present (i. e., unless the system is in free fall). When the natural convection is not negligible, such flows are typically referred to as mixed convection.

[11] When analysing potentially mixed convection, a parameter called the Archimedes number (Ar) parametizes the relative strength of free and forced convection. **The Archimedes number is the ratio of Grashof number and the square of Reynolds number, which represents the ratio of buoyancy force and inertia force, and which stands in for the contribution of natural convection. When Ar 1, natural convection dominates and when Ar 1, forced convection dominates.**

$$Ar = \frac{Gr}{Re^2} \tag{5}$$

[12] When natural convection isn't a significant factor, mathematical analysis with

forced convection theories typically yields accurate results. The parameter of importance in forced convection is the Peclet number (Pe), which is the ratio of advection (movement by currents) and diffusion (movement from high to low concentrations) of heat.

$$Pe = \frac{UL}{\alpha} \tag{6}$$

[13] As is common with fluid mechanics analysis, a number of dimensionless parameters are employed to describe convective heat transfer. A summary of these variables is included in the following tables:

Parameter	Formula	Interpretation
Prandtl Number:	$Pr = \dfrac{v}{\alpha} = \dfrac{c_p u}{k}$	Ratio of fluid velocity boundary layer thickness to the fluid temperature boundary layer thickness.
Nusselt Number:	$Nu = \dfrac{hL}{k}$	Ratio of heat transferred from surface to heat conducted away by fluid.
Reynolds Number:	$Re_L = \dfrac{u_\infty L}{v} = \dfrac{\rho u_\infty L}{\mu}$	Ratio of fluid inertia stress to viscous stress (for flow over flat plates).
Reynolds Number:	$Re_D = \dfrac{u_\infty D}{v} = \dfrac{\rho u_\infty D}{\mu}$	(Reynolds Number for pipe flow).
Stanton Number:	$St = \dfrac{h}{\rho c_p u_\infty} = \dfrac{Nu}{Re \cdot Pr}$	
Grashof Number:	$Gr = \dfrac{g\beta \Delta T L^3}{v^2}$	Ratio of fluid buoyancy stress to viscous stress.
Rayleigh Number:	$Ra = Gr \cdot Pr$	

Newton's Law of Cooling

$$\dot{Q} = hA(T_w - T_\infty) = hA \cdot \Delta T \tag{7}$$

[14] The rate of heat $\dot{Q}$ transfered to the surrounding fluid is proportional to the object's exposed area A, and the difference between the object temperature T_w and the fluid free-stream temperature T_∞.

Radiation

[15] Radiation heat transfer is concerned with the exchange of thermal radiation energy between two or more bodies. Thermal radiation is defined as electromagnetic radiation in the wavelength range of 0.1 to 100 microns (which encompasses the visible light regime), and arises as a result of a temperature difference between 2 bodies.

[16] No medium need exist between the two bodies for heat transfer to take place (as is needed by conduction and convection). Rather, the intermediaries are photons which travel at the speed of light. The heat transferred into or out of an object by thermal radiation is a function of several components. These include its surface reflectivity, emissivity,

surface area, temperature, and geometric orientation with respect to other thermally participating objects. In turn, an object's surface reflectivity and emissivity is a function of its surface conditions (roughness, finish, etc.) and composition.

Absorption and Emissivity

[17] Radiation heat transfer must account for both incoming and outgoing thermal radiation. Incoming radiation can be absorbed, reflected, or transmitted. This decomposition can be expressed by the relative fractions,

$$1 = \varepsilon_{reflected} + \varepsilon_{absorbed} + \varepsilon_{transmitted} \tag{8}$$

Since most solid bodies are opaque to thermal radiation, we can ignore the transmission component and write,

$$1 = \varepsilon_{reflected} + \varepsilon_{absorbed} \tag{9}$$

[18] **To account for a body's outgoing radiation (or its emissive power, defined as the heat flux per unit time), one makes a comparison to a perfect body who emits as much thermal radiation as possible.** Such an object is known as a blackbody, and the ratio of the actual emissive power E to the emissive power of a blackbody is defined as the surface emissivity,

$$\varepsilon = \frac{E}{E_{blackbody}} \tag{10}$$

[19] By stating that a body's surface emissivity is equal to its absorption fraction, Kirchhoff's Identity binds incoming and outgoing radiation into a useful dependent relationship,

$$\varepsilon = \varepsilon_{absorbed} \tag{11}$$

Black/Gray Bodies

[20] The heat emitted by a blackbody (per unit time) at an absolute temperature of T is given by the Stefan-Boltzmann Law of thermal radiation,

$$\dot{Q} = A\sigma T^4 = AE_{blackbody} \tag{12}$$

where $\dot{Q}$ has units of Watts, A is the total radiating area of the blackbody, and s is the Stefan-Boltzmann constant.

[21] A small blackbody at absolute temperature T enclosed by a much larger blackbody at absolute temperature T_e will transfer a net heat flow of,

$$\dot{Q} = A\sigma (T^4 - T_e^4) \tag{13}$$

[22] Bodies that emit less thermal radiation than a blackbody have surface emissivities less than 1. If the surface emissivity is independent of wavelength, then the body is called a "gray" body, in that no particular wavelength (or color) is favored.

[23] The net heat transfer from a small gray body at absolute temperature T with surface emissivity e to a much larger enclosing gray (or black) body at absolute temperature T_e is given by,

$$\dot{Q} = \varepsilon A\sigma (T^4 - T_e^4) \tag{14}$$

Radiation Angle Factors

[24] The above equations for blackbodies and graybodies assumed that the small body

could see only the large enclosing body and nothing else. Hence, all radiation leaving the small body would reach the large body.

[25] For the case where two objects can see more than just each other, then one must introduce a angle factor F and the heat transfer calculations become significantly more involved. The angle factor F_{12} is used to parameterize the fraction of thermal power leaving object 1 and reaching object 2. Specifically, this quantity is equal to,

$$\dot{Q}_{1\to 2} = A_1 F_{12} \varepsilon_1 \sigma T_1^4 \tag{15}$$

Likewise, the fraction of thermal power leaving object 2 and reaching object 1 is given by,

$$\dot{Q}_{2\to 1} = A_2 F_{21} \varepsilon_2 \sigma T_2^4 \tag{16}$$

[26] The case of two blackbodies in thermal equilibrium can be used to derive the following reciprocity relationship for angle factors,

$$A_1 F_{12} = A_2 F_{21} \tag{17}$$

Thus, once one knows F_{12}, F_{21} can be calculated immediately.

Heat Transfer Between Two Finite Graybodies

[27] The heat flow transferred from Object 1 to Object 2 where the two objects see only a fraction of each other and nothing else is given by,

$$\dot{Q} = \left(\frac{1-\varepsilon_1}{\varepsilon_1} + \frac{1}{F_{12}} + \left(\frac{1-\varepsilon_2}{\varepsilon_2}\right)\frac{A_1}{A_2}\right)^{-1} A_1 \sigma (T_1^4 - T_2^4) \tag{18}$$

[28] **This equation demonstrates the usage of F_{12}, but it represents a non-physical case since it would be impossible to position two finite objects such that they can see only a portion of each other and "nothing" else.** On the contrary, the complementary view factor $(1 - F_{12})$ cannot be neglected as radiation energy sent in those directions must be accounted for in the thermal bottom line. A more realistic problem would consider the same two objects surrounded by a third surface that can absorb and readmit thermal radiation yet is non-conducting. In this manner, all thermal energy that is absorbed by this third surface will be readmitted; no energy can be removed from the system through this surface. The equation describing the heat flow from Object 1 to Object 2 for this arrangement is,

$$\dot{Q} = \left(\frac{1-\varepsilon_1}{\varepsilon_1} + \frac{A_1 + A_2 - 2A_1 F_{12}}{A_2 - A_1 (F_{12})^2} + \left(\frac{1-\varepsilon_2}{\varepsilon_2}\right)\frac{A_1}{A_2}\right)^{-1} A_1 \sigma (T_1^4 - T_2^4) \tag{19}$$

Overall Resistance and Heat Transfer coefficient

[29] **In Equation (1) for conduction in a slab, Equation (12) for radiative heat transfer rate between two surfaces, and Equation (7) for convective heat transfer rate from a surface, the heat transfer rate is expressed as a temperature difference divided by a thermal resistance. Using the electrical resistance analogy, with temperature difference and heat transfer rate instead of potential difference and current, respectively, tools for solving series electrical resistance circuits can also be applied to heat transfer circuits.** For example, consider the heat transfer rate from a liquid to the surrounding gas separated by a constant cross-sectional area solid, as shown in Figure 1.

[30] The heat transfer rate from the fluid to the adjacent surface is by convection,

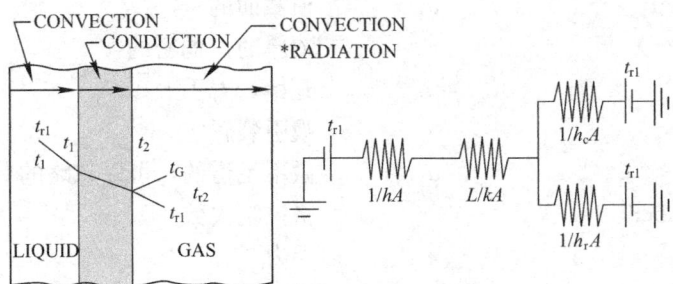

Fig. 1 Thermal Circuit

then across the solid body by conduction, and finally from the solid surface to the surroundings by both convection and radiation. A circuit using the equations for resistances in each mode is also shown. From the circuit, the heat transfer rate is

$$q = \frac{(t_{f1} - t_{f2})}{R_1 + R_2 + R_3} \tag{20}$$

where

$$R_1 = 1/hA \qquad R_2 = L/kA \qquad R_3 = \frac{(1/h_cA)(1/h_rA)}{(1/h_cA) + (1/h_cA)}$$

[31] Resistance R_3 is the parallel combination of the convection and radiation resistances on the right-hand surface, $1/h_cA$ and $1/h_rA$. Equivalently, $R_3 = 1/h_{rc}A$, where h_{rc} on the air side is the sum of the convection and radiation heat transfer coefficients (i.e. $h_{rc} = h_c + h_r$). The heat transfer rate can also be written as

$$q = UA(t_{f1} - t_{f2}) \tag{21}$$

where U is the overall heat transfer coefficient that accounts for all the resistances involved. Notethat

$$\frac{t_{f1} - t_{f2}}{q} = \frac{1}{UA} = R_1 + R_2 + R_3 \tag{22}$$

[32] The product UA is overall conductance, the reciprocal of overall resistance. The surface area A on which U is based is not always constant as in this example, and should always be specified when referring to U.

[33] Heat transfer rates are equal from the warm liquid to the solid surface, through the solid, and then to the cool gas. Temperature drops across each part of the heat flow path are related to the resis- tances (as voltage drops are in an electric circuit), so that

$$t_{f1} - t_1 = qR_1 \qquad t_1 - t_2 = qR_2 \qquad t_2 - t_{f2} = qR_3 \tag{23}$$

II. Words and Expressions

discipline	*n.*	学科
progressive	*a.*	连续的
electron	*n.*	电子
Fourier's Law		傅里叶定律

cross-sectional	*a.*	横截面的
dense	*a.*	浓厚的，稠密的
onset	*n.*	开始，着手
suction device		吸引装置
negligible	*a.*	微不足道的，可以忽略的
Grashof number		格拉晓夫数
buoyancy	*n.*	浮力
Peclet number		［化］佩克莱数
advection	*n.*	平流，水平对流
diffusion	*n.*	扩散，传播
dimensionless	*a.*	无量纲的，无因次的
proportional	*a.*	正比的，成比例的
electromagnetic	*a.*	电磁的
encompass	*v.*	包围，环绕
reflectivity	*n.*	反射率
emissivity	*n.*	发射率
geometric	*a.*	几何的，几何学的
orientation	*a.*	定向，定位，方位
opaque	*a.*	不透明的，不传热的
portion	*n.*	部分
slab	*n.*	平板
view factor	*n.*	角系数
tance	*n.*	铁磁性物质
reciprocal	*n.*	倒数
adjacent	*a.*	邻近的，毗邻的
analogy	*n.*	类似，类比

III. Notations

1. Heat transfer by conduction may be though of as the heat transferred through a substance (or combination of substance) from a region of high temperature to a region of low temperature by the progressive exchange of energy between the molecules of the substance.

由于传导而发生的热交换可认为是从单一物体（或复合体）的高温区到低温区的传热，这种传热是通过物体分子间连续的能量交换而实现的。

2. A common observation of convection is of thermal convection in a pot of boiling water, in which the hot and less-dense water on the bottom layer moves upwards, and the cool and more dense water near the top of the pot likewise sinks.

一个普遍的热对流现象是壶中沸水的热对流，壶底的热的、低密度的水上升，接近壶顶冷的、高密度的水下降。

3. The Archimedes number is the ratio of Grashof number and the square of Reynolds number, which represents the ratio of buoyancy force and inertia force, and which stands in for the contribution of natural convection. When $Ar \gg 1$, natural convection dominates and when $Ar \ll 1$, forced convection dominates.

stands in for 代替…

"and which stands in for the contribution of natural convection"这句话中的which指的是前文中的mixed convection

阿基米德数是格拉晓夫数和雷诺数平方的比值,表示着浮力和惯性力的比值,代表自然对流(对总流动)的贡献。当Ar远大于1时,自然对流占主导,当Ar远小于1时,强迫对流占主导地位。

4. To account for a body's outgoing radiation (or its emissive power, defined as the heat flux per unit time), one makes a comparison to a perfect body who emits as much thermal radiation as possible.

account for 解释,说明

为了解释物体的对外辐射(或物体的辐射力,被定义为每单位时间的热通量),物体要和理想物体进行对比较,理想物体的热辐射最大。

5. This equation demonstrates the usage of F_{12}, but it represents a non-physical case since it would be impossible to position two finite objects such that they can see only a portion of each other and "nothing" else.

这个方程解释了F_{12}的用法,但是没有这种实际情况,因为不可能安置两个确定的物体,它们只能看到对方的一部分而其他部分看不到。

6. In Equation (1) for conduction in a slab, Equation (12) for radiative heat transfer rate between two surfaces, and Equation (7) for convective heat transfer rate from a surface, the heat transfer rate is expressed as a temperature difference divided by a thermal resistance.

方程(1)是平板的热传导,方程(12)是两个表面的辐射热交换量,方程(7)是一个表面的热对流量,热交换量被表示为温差除以热阻。

7. Using the electrical resistance analogy, with temperature difference and heat transfer rate instead of potential difference and current, respectively, tools for solving series electrical resistance circuits can also be applied to heat transfer circuits.

使用电阻作类比,分别用温差和热交换量代替势差和电流,解决电阻环路的方法也可以应用于热交换环路。

IV. Exercises

1. Translate the following sentences into Chinese.

(1) When there exists a temperature gradient within a body, heat energy will flow from the region of high temperature to the region of low temperature. This phenomenon is known as conduction heat transfer, and is described by Fourier's Law.

(2) However, in any forced convection situation, some amount of natural convection

is always present whenever there are g-forces present (i.e., unless the system is in free fall). When the natural convection is not negligible, such flows are typically referred to as mixed convection.

(3) Radiation heat transfer is concerned with the exchange of thermal radiation energy between two or more bodies. Thermal radiation is defined as electromagnetic radiation in the wavelength range of 0.1 to 100 microns (which encompasses the visible light regime), and arises as a result of a temperature difference between 2 bodies.

(4) Bodies that emit less thermal radiation than a blackbody have surface emissivities less than 1. If the surface emissivity is independent of wavelength, then the body is called a "gray" body, in that no particular wavelength (or color) is favored.

(5) For the case where two objects can see more than just each other, then one must introduce a angle factor F and the heat transfer calculations become significantly more involved. The angle factor F_{12} is used to parameterize the fraction of thermal power leaving object 1 and reaching object 2.

(6) Heat transfer rates are equal from the warm liquid to the solid surface, through the solid, and then to the cool gas. Temperature drops across each part of the heat flow path are related to the resistances.

2. Translate the following sentences into English.

(1) 导热可以认为是通过物质分子间逐渐进行的能量交换而在物体内从高温区到低温区传递热量。在导热过程中，没有产生分子的具体的位移。然而，就金属来说，自由电子的运动大大有助于导热。

(2) 与表面直接接触的流体由于导热而被加热，变轻，并且由于与相邻流体的密度差而上升。这种运动由于流体黏性而受到阻碍。热传递受下列因素影响：1) 因热膨胀而引起的重力作用；2) 黏性阻滞；3) 热扩散。这种热传递被认为取决于重力加速度、热膨胀系数、运动黏滞系数及导热系数。

(3) 对不同几何形状、辐射特性和方向的表面，估计它们之间的传热速率时，常假定：1) 所有表面均是灰的或黑的；2) 辐射和反射式漫射；3) 整个表面参数均匀一致；4) 吸收率等于发射率并且与投射辐射源的温度无关；5) 两辐射表面间的物质既不发射也不吸收辐射。这些假设由于它们提供的简化程度很大而被使用，虽然得出的结果只能认为是近似的。

Lesson 4 Fluid Mechanics (流体力学)

I. Text

[1] Fluid mechanics, branch of mechanics dealing with the properties and behavior of fluids, i. e., liquids and gases. Because of their ability to flow, liquids and gases have many properties in common not shared by solids. The special study of fluids in motion, or fluid dynamics, makes up the larger part of fluid mechanics. Branches of fluid dynamics include hydrodynamics (study of liquids in motion) and aerodynamics (study of gases in motion). Hydrodynamics is often used synonymously with fluid dynamics, since most of the results from the study of liquids also apply to gases. The study of plasmas in motion is known as magneto hydrodynamic and includes principles from several fields.

Fluid Preliminaries

[2] By definition, a fluid is a material continuum that is unable to withstand a static shear stress. Unlike an elastic solid which responds to a shear stress with a recoverable deformation, a fluid responds with an irrecoverable flow.

Variables needed to define a fluid and its environment is:

Quantity	Symbol	Object	Units
pressure	p	scalar	N/m^2
velocity	v	vector	m/s
density	ρ	scalar	kg/m^3
viscosity	μ	scalar	$kg/(m \cdot s)$
body force	b	vector	N/kg
time	t	scalar	s
kinematic viscosity	$\upsilon = \mu/\rho$	scalar	cm^2/s

[3] **Both liquids and gases are fluids, although the natures of their molecular interactions differ strongly in both degree of compressibility and formation of a free surface (interface) in liquid.** In general, liquids are considered incompressible fluids; gases may range from compressible to nearly incompressible. Liquids have unbalanced molecular cohesive forces at or near the surface (interface), so the liquid surface tends to contract and has properties similar to a stretched elastic membrane. A liquid surface, therefore, is under tension (surface tension).

[4] Fluid motion can be described by several simplified models. The simplest is the ideal-fluid model, which assumes that the fluid has no resistance to shearing. Ideal fluid flow analysis is well developed (e. g., Schlichting 1979), and may be valid for a wide range

of applications.

Types of Flow

[5] Fluid flow can be either laminar or turbulent. Laminar and turbulent flows can be differentiated using the Reynolds number Re, which is a dimensionless relative ratio of inertial forces to viscous forces:

$$Re_L = vL/\upsilon \tag{1}$$

where L is the characteristic length scale and υ is the kinematic viscosity of the fluid. In flow through pipes, tubes, and ducts, the characteristic length scale is the hydraulic diameter D, given by

$$D_h = 4A/P_w \tag{2}$$

where A is the cross-sectional area of the pipe, duct, or tube, and P_w is the wetted perimeter.

[6] For a round pipe, D_h equals the pipe diameter. In general, laminar low in pipes or ducts exists when the Reynolds number (based on D_h) is less than 2300. Fully turbulent flow exists when $Re_{Dh} > 10\ 000$. For $2300 < Re_{Dh} < 10\ 000$, transitional flow exists, and predictions are unreliable.

Laminar Flow

[7] **When real-fluid effects of viscosity or turbulence are included, the continuity relation is not changed, but v must be evaluated from the integral of the velocity profile, using local velocities.** In fluid flow past fixed boundaries, velocity at the boundary is zero, velocity gradients exist, and shear stresses are produced. The equations of motion then become complex, and exact solutions are difficult to find except in simple cases for laminar flow between flat plates, between rotating cylinders, or within a pipe or tube.

[8] For steady, fully developed laminar flow between two parallel plates (Figure 1), shear stress varies linearly with distance y from the centerline (transverse to the flow; $y = 0$ in the center of the channel). For a wide rectangular channel $2b$ tall, can be written as

$$\tau = \left(\frac{y}{b}\right)\tau_w = \mu \frac{d\upsilon}{dy} \tag{3}$$

where τ_w is wall shear stress $[b\ (dp/ds)]$, and s is flow direction. Because velocity is zero at the wall ($y=b$), Equation (3) can be integrated to yield

$$\upsilon = \left(\frac{b^2 - y^2}{2\mu}\right)\frac{dp}{ds} \tag{4}$$

[9] The resulting parabolic velocity profile in a wide rectangular channel is commonly called Poiseuille flow. Maximum velocity occurs at the centerline ($y = 0$), and the average velocity v is 2/3 of the maximum velocity. From this, the longitudinal pressure drop in terms of v can be written as

$$\frac{dp}{ds} = -\left(\frac{3\mu\upsilon}{R^2}\right) \tag{5}$$

[10] A parabolic velocity profile can also be derived for a pipe of radius R. υ is 1/2 of

the maximum velocity, and the pressure drop can be written as

$$\frac{\mathrm{d}p}{\mathrm{d}s} = -\left(\frac{8\mu\omega}{R^2}\right) \tag{6}$$

Turbulence Flow

[11] Fluid flows are generally turbulent, involving random perturbations or fluctuations of the flow (velocity and pressure), characterized by an extensive hierarchy of scales or frequencies (Robertson 1963). **Flow disturbances that are not chaotic but have some degree of periodicity (e. g. , the oscillating vortex trail behind bodies) have been erroneously identified as turbulence.** Only flows involving ran-

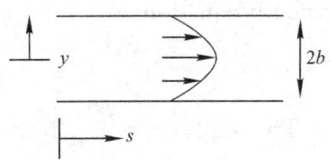

Fig. 1 Dimensions for Steady, Fully Developed Laminar Flow Equations

dom perturbations without any order or periodicity are turbulent; velocity in such a flow varies with time or locale of measurementc Figure 2).

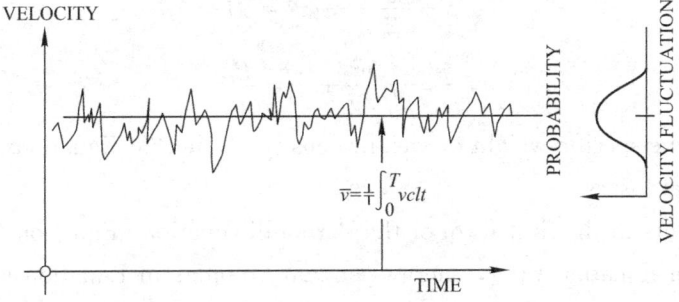

Fig. 2 Velocity Fluctuation at Point in Turbulent Flow

[12] Turbulence can be quantified statistically. The velocity most often used is the time-averaged velocity. The strength of turbulence is characterized by the root mean square (RMS) of the instantaneous variation in velocity about this mean. Turbulence causes the fluid to transfer momentum, heat, and mass very rapidly across the flow.

Bernoulli Equation

[13] The Bernoulli equation is a fundamental principle of fluid flow analysis. It involves the conservation of momentum and energy along a streamline; it is not generally applicable across streamlines. Development is fairly straightforward. The first law of thermodynamics can apply to both mechanical flow energies (kinetic and potential energy) and thermal energies.

[14] The change in energy content ΔE per unit mass of flowing fluid is a result of the work per unit mass w done on the system plus the heat per unit mass q absorbed or rejected:

$$\Delta E = w + q \tag{7}$$

Fluid energy is composed of kinetic, potential (because of elevation z), and internal (u) energies. Per unit mass of fluid, the energy change relation between two sections of the system is

$$\Delta\left(\frac{v^2}{2} + gz + u\right) = E_M - \Delta\left(\frac{p}{\rho}\right) + q \tag{8}$$

where the work terms are (1) external work E_M from a fluid machine (E_M is positive for a pump or blower) and (2) flow work p/ρ (where p = pressure), and g is the gravitational constant. Rearranging, the energy equation can be written as the generalized Bernoulli equation:

$$\Delta\left(\frac{v^2}{2}+gz+u+\frac{p}{\rho}\right)=E_M+q \tag{9}$$

The expression in parentheses in Equation (9) is the sum of the kinetic energy, potential energy, internal energy, and flow work per unit mass flow rate. **In cases with no work interaction, no heat transfer, and no viscous frictional forces that convert mechanical energy into internal energy, this expression is constant and is known as the Bernoulli constant B:**

$$\frac{v^2}{2}+gz+\left(\frac{p}{\rho}\right)=B \tag{10}$$

Alternative forms of this relation are obtained through multiplication by ρ or division by g:

$$p+\frac{\rho v^2}{2}+\rho gz=\rho B \tag{11}$$

$$\frac{p}{\gamma}+\rho\frac{v^2}{2g}+z=\frac{B}{g} \tag{12}$$

where $\gamma=\rho g$ is the specific weight or weight density. Note that Equations (10) to (12) assume no frictional losses.

[15] The units in the first form of the Bernoulli equation [Equation (10)] are energy per unit mass; in Equation (10), energy per unit volume; in Equation (12), energy per unit weight, usually called head. Note that the unit for head reduce is just length. In gas flow analysis, Equation (11) is often used, and ρgz is negligible. Equation (11) should be used when density variations occur. For liquid flows, Equation (12) is commonly used. Identical results are obtained with the three forms if the units are consistent and fluids are homogeneous.

[16] Many systems of pipes, ducts, pumps, and blowers can be considered as one-dimensional flow along a streamline (i. e. , variation in velocity across the pipe or duct is ignored, and local velocity v = average velocity V). When v varies significantly across the cross section, the kinetic energy term in the Bernoulli constant B is expressed as $\alpha v^2/2$, where the kinetic energy factor ($\alpha>1$) expresses the ratio of the true kinetic energy of the velocity profile to that of the average velocity. For laminar flow in a wide rectangular channel, $\alpha=1.54$, and in a pipe, $\alpha=2.0$. For turbulent flow in a duct, $\alpha\approx1$.

[17] Heat transfer q may often be ignored. Conversion of mechanical energy to internal energy Δu may be expressed as a loss E_L. The change in the Bernoulli constant ($\Delta B = B_2 - B_1$) between stations 1 and 2 along the conduit can be expressed as

$$\left(\frac{p}{\rho}+\alpha\frac{V^2}{2}+gz\right)_1+E_M-E_L=\left(\frac{p}{\rho}+\alpha\frac{V^2}{2}+gz\right)_2 \tag{13}$$

or, by dividing by g, in the form

$$\left(\frac{p}{\gamma}+\alpha\frac{V^2}{2g}+z\right)_1+H_M-H_L=\left(\frac{p}{\gamma}+\alpha\frac{V^2}{2G}+z\right)_2 \tag{14}$$

[18] Note that Equation (13) has units of energy per mass, whereas each term in Equation (14) has units of energy per weight, or head. The terms E_M and E_L are defined as positive, where $H_M = E_M$ represents energy added to the conduit flow by pumps or blowers. A turbine or fluid motor thus has a negative H_M or E_M. The terms E_M and H_M (= E_M/g) are defined as positive, and represent energy added to the fluid by pumps or blowers. The simplicity of Equation (14) should be noted; the total head at station 1 (pressure head plus velocity head plus elevation head) plus the head added by a pump (H_M) minus the head lost through friction (H_L) is the total head at station 2.

Compressible and Incompressible Fluids

[19] Fluid mechanics deals with both incompressible and compressible fluids, that is, with fluids of either constant or variable density. Although there is no such thing in reality as an incompressible fluid, this is usually the case with liquids. Gases, too, may be considered incompressible when the pressure variation is small compared with the absolute pressure.

[20] Liquids are ordinary considered incompressible fluids. But sound waves, which are really pressure waves, travel through them. This is evidence of the elasticity of liquids. In problems involving water hammer, it is necessary to consider the compressible of the liquid.

[21] The flow of air in a ventilating system is a case. There a gas may be treated as incompressible, for the pressure variation is so small that the change in density is of no importance. But for a gas or stream flowing at high velocity through a long pipeline, the drop in pressure may be so great that change in density cannot be ignored. For an airplane flying at speeds below 100 m/s, the air may be considered to be of constant density. **But as an object moving through the air approaches the velocity of sound, which is of the order of 300 m/s, the pressure and density of the air adjacent to the body become materially different from those of the air at some distance away. And the air must then be treated as a compressible fluid.**

[22] The compressible of a liquid is inversely proportional to its volume modulus of elasticity, also known as the bulk modulus. This modulus is defined as, $E_v = -vdp/dv = -(v/dv)dp$, where $v =$ specific volume and $p =$ unit pressure. As v/dv is a dimensionless ration, the units of E_v and p are the same. The bulk modulus is analogous to the modulus of elasticity for solids. However, for fluids it is defined on a volume basis rather than in terms of the familiar one-dimensional stress-strain relation for solid bodies.

Basic Flow Processes

Wall Friction

[23] Laminar and turbulent flow differs significantly in their velocity profiles. Turbulent flow profiles are flat and laminar profiles are more pointed (Figure 3). **As discussed, fluid velocities of the turbulent profile near the wall must drop to zero more rapidly than those of the laminar profile, so shear stress and friction are much greater in turbulent flow.** Fully

developed conduit flow may be characterized by the pipe factor, which is the ratio of average to maximum (centerline) velocity. Viscous velocity profiles result in pipe factors of 0.667 and 0.50 for wide rectangular and axisymmetric conduits.

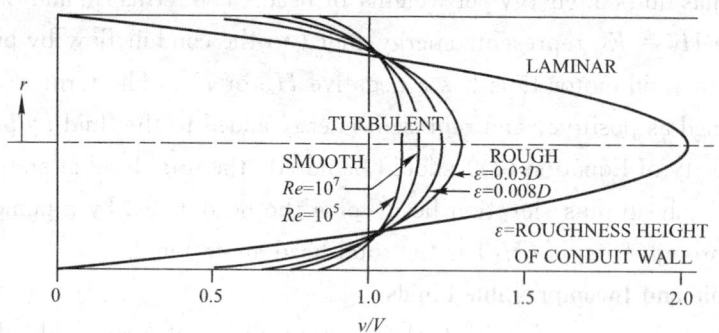

Fig. 3 Velocity Profiles of Flow in Pipes

Boundary Layer

[24] The boundary layer is the region close to the wall where wall friction affects flow. Boundary layer thickness (usually denoted by δ) is thin compared to downstream flow distance. For external flow over a body, fluid velocity varies from zero at the wall to a maximum at distance δ from the wall. Boundary layers are generally laminar near the start of their formation but may become turbulent downstream.

[25] A significant boundary-layer occurrence exists in a pipeline or conduit following a well-rounded entrance (Figure 4). Layers grows from the walls until they meet at the center of the pipe. Near the start of the straight conduit, the layer is very thin and most likely laminar, so the uniform velocity core outside has a velocity only slightly greater than the average velocity. **As the layer grows in thickness, the slower velocity near the wall requires a velocity increase in the uniform core to satisfy continuity. As the flow proceeds, the wall layers grow (and the centerline velocity increases) until they join, after an entrance length L_e.** Applying the Bernoulli relation of Equation (11) to core flow indicates a decrease in pressure along the layer.

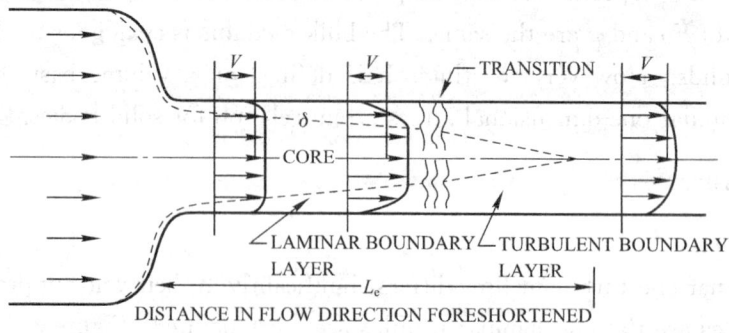

Fig. 4 Flow in Conduit Entrance Region

II. Words and Expressions

dynamics	n.	动力学，力学
hydrodynamics	n.	流体力学，流体动力学
aerodynamics	n.	气体力学
plasma	n.	等离子体
magneto	n.	磁发电机
hear stress		剪应力，剪切应力
scalar	n.	标量
vector	n.	矢量
deformation	n.	变形
body force		质量力
compressibility	n.	压缩性，压缩系数
interface	n.	接口，接触面
membrane	n.	膜，薄膜
tension	n.	张力，拉力
inertial force		惯性力
viscous force		黏性力
hydraulic	a.	水力的，水力学的
wetted perimeter		湿周
integrated	a.	综合的，完整的
parabolic	a.	抛物线的
longitudinal	a.	经度的，纵向的
conduit	n.	管道，导管
profile	n.	剖面，轮廓，分布图
perturbation	n.	扰动，微扰，波动
chaotic	a.	混乱的，无序的
periodicity	n.	周期
homogeneous	a.	同类的，相似的，均匀的
blower	n.	鼓风机，排风机
elasticity	n.	弹性，弹力
duct	n.	管，输送管，排泄管
	v.	用管道输送
axisymmetric	a.	轴对称的
in terms of		在…方面，依据
hammer	v.	锤击，捶打

III. Notations

1. Both liquids and gases are fluids, although the natures of their molecular interac-

tions differ strongly in both degree of compressibility and formation of a free surface (interface) in liquid.

虽然液体和气体分子的界面性质在压缩程度和自由液面的形成这两个方面有很大差异，但是它们都属于流体。

2. When real-fluid effects of viscosity or turbulence are included, the continuity relation is not changed, but v must be evaluated from the integral of the velocity profile, using local velocities.

当包含真实流体的黏性和扰动影响时，连续性关系式不会改变，但是 v 必须对局部速度的分布进行积分计算获得。

3. Flow disturbances that are not chaotic but have some degree of periodicity (e. g., the oscillating vortex trail behind bodies) have been erroneously identified as turbulence.

some degree of　　一定程度的

流体扰动不是无序的，而是具有一定程度的周期性（例如，物体之后的振动涡流线），这些扰动被错误的确定为湍流。

4. In cases with no work interaction, no heat transfer, and no viscous frictional forces that convert mechanical energy into internal energy, this expression is constant and is known as the Bernoulli constant B.

在没有功量的相互作用、没有热交换、没有黏性摩擦力的情况下，把机械能转化为内能的表达式是恒定的，这就是著名的伯努利方程。

5. But as an object moving through the air approaches the velocity of sound, which is of the order of 300 m/s, the pressure and density of the air adjacent to the body become materially different from those of the air at some distance away. And the air must then be treated as a compressible fluid.

of the order of　　大约，达到…程度　　at some distance　　有一段距离

然而，当物体在空气中的速度接近声速，即大约是 300 m/s，毗邻物体的空气的压力和密度与稍远处空气的压力和密度实际上有很大差异。此时空气必须被视为可压缩流体。

6. As discussed, fluid velocities of the turbulent profile near the wall must drop to zero more rapidly than those of the laminar profile, so shear stress and friction are much greater in turbulent flow.

正如讨论的那样，接近墙体的湍流流体速度比层流更快的降低到零，所以，湍流的剪切应力和摩擦力更大。

7. As the layer grows in thickness, the slower velocity near the wall requires a velocity increase in the uniform core to satisfy continuity. As the flow proceeds, the wall layers grow (and the centerline velocity increases) until they join, after an entrance length L_e.

随着边界层厚度的增加，为了保持连续性，靠近壁面越低的流体速度需要均匀地增加。随着流动的进行，直到进口段长度 L_e 后，边界层和中心线相交。

IV. Exercises

1. Translate the following sentences into Chinese.

(1) Branches of fluid dynamics include hydrodynamics (study of liquids in motion) and aerodynamics (study of gases in motion). Hydrodynamics is often used synonymously with fluid dynamics, since most of the results from the study of liquids also apply to gases.

(2) Fluid flow can be either laminar or turbulent. Laminar and turbulent flows can be differentiated using the Reynolds number Re, which is a dimensionless relative ratio of inertial forces to viscous forces.

(3) Turbulence can be quantified statistically. The velocity most often used is the time-averaged velocity. The strength of turbulence is characterized by the root mean square (RMS) of the instantaneous variation in velocity about this mean. Turbulence causes the fluid to transfer momentum, heat, and mass very rapidly across the flow.

(4) Note that the units for head reduce to just length. In gas flow analysis, Equation (11) is often used, and $\rho g z$ is negligible. Equation (11) should be used when density variations occur. For liquid flows, Equation (12) is commonly used. Identical results are obtained with the three forms if the units are consistent and fluids are homogeneous.

(5) Fluid mechanics deals with both incompressible and compressible fluids, that is, with fluids of either constant or variable density. Although there is no such thing in reality as an incompressible fluid, this is usually the case with liquids. Gases, too, may be considered incompressible when the pressure variation is small compared with the absolute pressure.

(6) For external flow over a body, fluid velocity varies from zero at the wall to a maximum at distance δ from the wall. Boundary layers are generally laminar near the start of their formation but may become turbulent downstream.

2. Translate the following sentences into English.

(1) 流体动力学研究的主要问题是流速和压强在空间的分布，两者之间，流速又更加重要，与流速密切相关的有惯性力和黏性力。其中，惯性力是由质点本身流速变化所产生，而黏性力是由于流层与流层之间，质点与质点间存在着流速差异所引起的。

(2) 水和空气的黏滞系数随温度变化的规律是不同的，水的黏滞性随温度升高而减少，空气的黏滞性随温度升高而增大。这是因为黏滞性是分子间的吸引力和分子不规则的热运动产生动量交换的结果。温度升高，分子间吸引力降低，动量增大；反正，温度降低，分子间吸引力增大，动量减少。

(3) 由于密度差和重力作用引起流体运动而产生的热传递称为自然对流。自然对流的传热系数一般低于强迫对流，因此在计算总的得热量或失热量时，主要的一点就是不要忽略辐射热。辐射热传递与自然对流可以有相同的数量级，甚至在室温下也如此，因为室内墙体温度影响人体的舒适感。

Extensive Reading

History of Thermodynamics

[1] Basic physical notions of heat and temperature were established in the 1600s, and scientists of the time appear to have thought correctly that heat is associated with the mo-

tion of microscopic constituents of matter. But in the 1700s it became widely believed that heat was instead a separate fluid-like substance. Experiments by James Joule and others in the 1840s put this in doubt, and finally in the 1850s it became accepted that heat is in fact a form of energy. The relation between heat and energy was important for the development of steam engines, and in 1824 Sadi Carnot had captured some of the ideas of thermodynamics in his discussion of the efficiency of an idealized engine. Around 1850 Rudolf Clausius and William Thomson (Kelvin) stated both the First Law - that total energy is conserved - and the Second Law of Thermodynamics. The Second Law was originally formulated in terms of the fact that heat does not spontaneously flow from a colder body to a hotter. Other formulations followed quickly, and Kelvin in particular understood some of the law's general implications. The idea that gases consist of molecules in motion had been discussed in some detail by Daniel Bernoulli in 1738, but had fallen out of favor, and was revived by Clausius in 1857. Following this, James Clerk Maxwell in 1860 derived from the mechanics of individual molecular collisions the expected distribution of molecular speeds in a gas. Over the next several years the kinetic theory of gases developed rapidly, and many macroscopic properties of gases in equilibrium were computed. In 1872 Ludwig Boltzmann constructed an equation that he thought could describe the detailed time development of a gas, whether in equilibrium or not. In the 1860s Clausius had introduced entropy as a ratio of heat to temperature, and had stated the Second Law in terms of the increase of this quantity. Boltzmann then showed that his equation implied the so-called H Theorem, which states that a quantity equal to entropy in equilibrium must always increase with time. At first, it seemed that Boltzmann had successfully proved the Second Law. But then it was noticed that since molecular collisions were assumed reversible, his derivation could be run in reverse, and would then imply the opposite of the Second Law. Much later it was realized that Boltzmann's original equation implicitly assumed that molecules are uncorrelated before each collision, but not afterwards, thereby introducing a fundamental asymmetry in time. Early in the 1870s Maxwell and Kelvin appear to have already understood that the Second Law could not formally be derived from microscopic physics, but must somehow be a consequence of human inability to track large numbers of molecules. In responding to objections concerning reversibility Boltzmann realized around 1876 that in a gas there are many more states that seem random than seem orderly. This realization led him to argue that entropy must be proportional to the logarithm of the number of possible states of a system, and to formulate ideas about ergodicity.

[2] The statistical mechanics of systems of particles was put in a more general context by Willard Gibbs, beginning around 1900. Gibbs introduced the notion of an ensemble - a collection of many possible states of a system, each assigned a certain probability. He argued that if the time evolution of a single state were to visit all other states in the ensemble - the so-called ergodic hypothesis - then averaged over a sufficiently long time a single state would behave in a way that was typical of the ensemble. Gibbs also gave qualitative argu-

ments that entropy would increase if it was measured in a "coarse-grained" way in which nearby states was not distinguished. In the early 1900s the development of thermodynamics was largely overshadowed by quantum theory and little fundamental work was done on it. Nevertheless, by the 1930s, the Second Law had somehow come to be generally regarded as a principle of physics whose foundations should be questioned only as a curiosity. Despite neglect in physics, however, ergodic theory became an active area of pure mathematics, and from the 1920s to the 1960s properties related to ergodicity were established for many kinds of simple systems. When electronic computers became available in the 1950s, Enrico Fermi and others began to investigate the ergodic properties of nonlinear systems of springs. But they ended up concentrating on recurrence phenomena related to solitons, and not looking at general questions related to the Second Law. Much the same happened in the 1960s, when the first simulations of hard sphere gases were led to concentrate on the specific phenomenon of long-time tails.

[3] And by the 1970s, computer experiments were mostly oriented towards ordinary differential equations and strange attractors, rather than towards systems with large numbers of components, to which the Second Law might apply. Starting in the 1950s, it was recognized that entropy is simply the negative of the information quantity introduced in the 1940s by Claude Shannon. Following statements by John von Neumann, it was thought that any computational process must necessarily increase entropy, but by the early 1970s, notably with work by Charles Bennett, it became accepted that this is not so, laying some early groundwork for relating computational and thermodynamic ideas.

A Brief History of Computational Fluid Dynamics (CFD)

[1] Since the dawn of civilization, mankind has always had a fascination with fluids; whether it is the flow of water in rivers, the wind and weather in our atmosphere, the smelting of metals, powerful ocean currents or the flow of blood around our bodies.

[2] In antiquity, great Greek thinkers like Heraclitus postulated that "Everything flows" but he was thinking of this in a philosophical sense rather than in a recognizably scientific way. However, Archimedes initiated the fields of static mechanics, hydrostatics, and determined how to measure densities and volumes of objects. The focus at the time was on waterworks: aqueducts, canals, harbors, and bathhouses, which the ancient Romans perfected to a science.

[3] It was not until the Renaissance that these ideas resurfaced again in Southern Europe when we find great artists cum engineers like Leonardo Da Vinci starting to examine the natural world of fluids and flow in detail again. He observed natural phenomena in the visible world, recognizing their form and structure, and describing them pictorially exactly as they were. He planned and supervised canal and harbor works over a large part of middle Italy. His contributions to fluid mechanics are presented in a nine part treatise (Del moto e misura dell'acqua) that covers water surfaces, movement of water, water waves,

eddies, falling water, free jets, interference of waves, and many other newly observed phenomena.

[4] Leonardo was followed in the late 17th Century by Isaac Newton in England. Newton tried to quantify and predict fluid flow phenomena through his elementary Newtonian physical equations. His contributions to fluid mechanics included his second law: $F=M \cdot a$, the concept of Newtonian viscosity in which stress and the rate of strain vary linearly, the reciprocity principle: the force applied upon a stationary object by a moving fluid is equal to the change in momentum of the fluid as it deflects around the front of the object, and the relationship between the speed of waves at a liquid surface and their wavelength.

[5] In the 18th and 19th centuries, significant work was done trying to mathematically describe the motion of fluids. Daniel Bernoulli (1700-1782) derived Bernoulli's famous equation, and Leonhard Euler (1707-1783) proposed the Euler equations, which describe the conservation of momentum for an inviscid fluid, and conservation of mass. He also proposed the velocity potential theory. Two other very important contributors to the field of fluid flow emerged at this time: the Frenchman, Claude Louis Marie Henry Navier (1785-1836) and the Irishman, George Gabriel Stokes (1819-1903) who introduced viscous transport into the Euler equations, which resulted in the now famous Navier-Stokes equation. These forms of the differential mathematical equations that they proposed nearly 200 years ago are the basis of the modern day computational fluid dynamics (CFD) industry, and they include expressions for the conservation of mass, momentum, pressure, species and turbulence. Indeed, the equations are so closely coupled and difficult to solve that it was not until the advent of modern digital computers in the 1960s and 1970s that they could be resolved for real flow problems within reasonable timescales. Other key figures who developed theories related to fluid flow in the 19th century were Jean Le Rond d'Alembert, Siméon-Denis Poisson, Joseph Louis Lagrange, Jean Louis Marie Poiseuille, John William Rayleigh, M. Maurice Couette, Osborne Reynolds, and Pierre Simon de Laplace.

[6] In the early 20th Century, much work was done on refining theories of boundary layers and turbulence in fluid flow. Ludwig Prandtl (1875-1953) proposed a boundary layer theory, the mixing length concept, compressible flows, the Prandtl number, and much more that we take for granted today. Theodore von Karman (1881-1963) analyzed what is now known as the von Karman vortex street. Geoffrey Ingram Taylor (1886-1975) proposed a statistical theory of turbulence and the Taylor microscale. Andrey Nikolaevich Kolmogorov (1903-1987) introduced the concept of Kolmogorov scales and the universal energy spectrum for turbulence, and George Keith Batchelor (1920-2000) made contributions to the theory of homogeneous turbulence.

[7] It is debatable as to who did the earliest CFD calculations (in a modern sense) although Lewis Fry Richardson in England (1881-1953) developed the first numerical weather prediction system when he divided physical space into grid cells and used the finite difference approximations of Bjerknes's "primitive differential equations". His own at-

tempt to calculate weather for a single eight-hour period took six weeks of real time and ended in failure! His model's enormous calculation requirements led Richardson to propose a solution he called the "forecast-factory". The "factory" would have involved filling a vast stadium with 64,000 people. Each one, armed with a mechanical calculator, would perform part of the flow calculation. A leader in the center, using colored signal lights and telegraph communication, would coordinate the forecast. What he was proposing would have been a very rudimentary CFD calculation. The earliest numerical solution for flow past a cylinder was carried out in 1933 by Thom and reported in England:

A. Thom, 'The Flow Past Circular Cylinders at Low Speeds', Proc. Royal Society, A141, pp. 651-666, London, 1933

Kawaguti in Japan obtained a similar solution for flow around a cylinder in 1953 by using a mechanical desk calculator, working 20 hours per week for 18 months.

[8] M. Kawaguti, 'Numerical Solution of the NS Equations for the Flow Around a Circular Cylinder at Reynolds Number 40', more details can be found in Journal of Phy. Soc. Japan, Vol. 8, pp. 747-757, 1953.

[9] During the 1960s, the theoretical division of NASA at Los Alamos in the U.S. contributed many numerical methods that are still in use in CFD today, such as the following methods: Particle-In-Cell (PIC), Marker-and-Cell (MAC), Vorticity-Stream function methods, Arbitrary Lagrangian-Eulerian (ALE) methods, and the ubiquitous k - e turbulence model. In the 1970s, a group working under D. Brian Spalding, at Imperial College, London, developed Parabolic flow codes (GENMIX), Vorticity-Stream function based codes, the SIMPLE algorithm and the TEACH code, as well as the form of the k - e equations that are used today (Spalding & Launder, 1972). They went on to develop Upwind differencing, 'Eddy break-up' and 'presumed pdf' combustion models. Another key event in CFD industry was in 1980 when Suhas V. Patankar published "Numerical Heat Transfer and Fluid Flow", probably the most influential book on CFD to date, and the one that spawned a thousand CFD codes.

[10] It was in the early 1980s that commercial CFD codes came into the open market place in a big way. The use of commercial CFD software started to become accepted by major companies around the world rather than their continuing to develop in-house CFD codes. Commercial CFD software is therefore based on sets of very complex non-linear mathematical expressions that define the fundamental equations of fluid flow, heat and materials transport. These equations are solved iteratively using complex computer algorithms embedded within CFD software. The net effect of such software is to allow the user to computationally model any flow field provided the geometry of the object being modeled is known, the physics and chemistry are identified, and some initial flow conditions are prescribed. Outputs from CFD software can be viewed graphically in color plots of velocity vectors, contours of pressure, lines of constant flow field properties, or as "hard" numerical data and X-Y plots.

[11] CFD is now recognized to be a part of the computer-aided engineering (CAE) spectrum of tools used extensively today in all industries, and its approach to modeling fluid flow phenomena allows equipment designers and technical analysts to have the power of a virtual wind tunnel on their desktop computer. CFD software has evolved far beyond what Navier, Stokes or Da Vinci could ever have imagined. CFD has become an indispensable part of the aerodynamic and hydrodynamic design process for planes, trains, automobiles, rockets, ships, submarines; and indeed any moving craft or manufacturing process that mankind has devised.

Lesson 5 Heating Engineering (供暖工程)

I. Text

[1] A heating system is a mechanism for maintaining temperatures at an acceptable level by using thermal energy within a home, office, or other dwelling. The earliest method of providing interior heating was an open fire. **Such a source, along with related methods such as fireplaces, cast-iron stoves, and modern space heaters fueled by gas or electricity, is known as direct heating because the conversion of energy into heat takes place at the site to be heated.** A more common form of heating in modern times is known as central, or indirect, heating. It consists of the conversion of energy to heat at a source outside of, apart from, or located within the site or sites to be heated; the resulting heat is conveyed to the site through a fluid medium such as air, water, or steam.

Heat Exchanger

[2] A heat exchanger is a device built for efficient heat transfer from one medium to another. The medium may be separated by a solid wall, so that they never mix, or they may be in direct contact. They are widely used in space heating, refrigeration, air conditioning, power plants, chemical plants, petrochemical plants, petroleum refineries, and natural gas processing.

Flow Arrangement

[3] Heat exchangers may be classified according to their flow arrangement (Fig. 1~Fig. 3). In parallel-flow heat exchangers, the two fluids enter the exchanger at the same end, and travel in parallel to one another to the other side. In counter-flow heat exchangers the fluids enter the exchanger from opposite ends. The counter current design is most

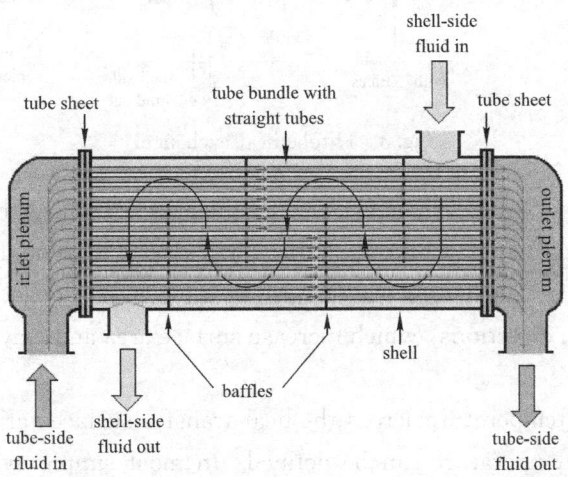

Fig. 1 Straight-tube heat exchanger (one pass tube-side)

efficient, in that it can transfer the most heat from the heat (transfer) medium. See countercurrent exchange. In a cross-flow heat exchanger, the fluids travel roughly perpendicular to one another through the exchanger.

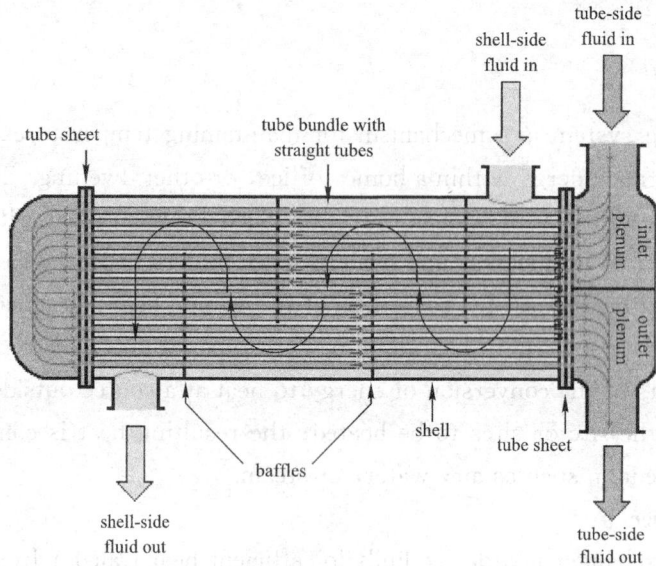

Fig. 2　Straight-tube heat enchanger (two pass tube-side)

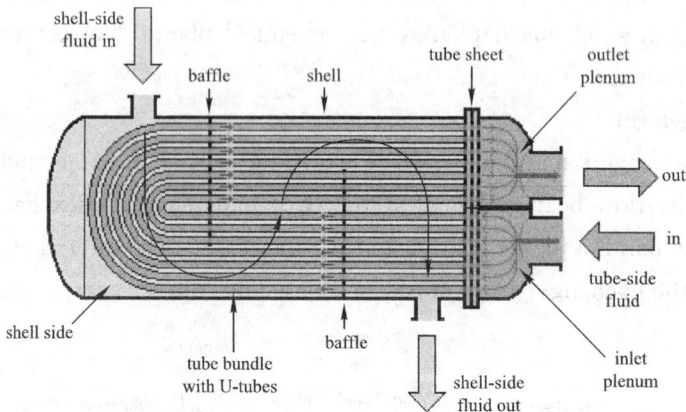

Fig. 3　U-tube heat exchanger

[4] For efficiency, heat exchangers are designed to maximize the surface area of the wall between the two fluids, while minimizing resistance to fluid flow through the exchanger. The exchanger's performance can also be affected by the addition of fins or corrugations in one or both directions, which increase surface area and may channel fluid flow or induce turbulence.

[5] The driving temperature across the heat transfer surface varies with position, but an appropriate meantemperature can be defined. In most simple systems this is the log mean temperature difference (LMTD). Sometimes direct knowledge of the LMTD is not

available and the NTU method is used.

Types of Heat Exchangers

- **Shell and Tube Heat Exchanger**

[6] Shell and tube heat exchangers consist of a series of tubes. One set of these tubes contains the fluid that must be either heated or cooled. The second fluid runs over the tubes that are being heated or cooled so that it can either provide the heat or absorb the heat required. A set of tubes is called the tube bundle and can be made up of several types of tubes: plain, longitudinally finned, etc. Shell and Tube heat exchangers are typically used for high pressure applications (with pressures greater than 30 bar and temperatures greater than 260℃). This is because the shell and tube heat exchangers are robust due to their shape.

- **Plate Heat Exchanger**

[7] **Another type of heat exchanger is the plate heat exchanger. One is composed of multiple, thin, slightly-separated plates that have very large surface areas and fluid flow passages for heat transfer. This stacked-plate arrangement can be more effective, in a given space, than the shell and tube heat exchanger.** Advances in gasket and brazing technology have made the plate-type heat exchanger increasingly practical. In HVAC applications, large heat exchangers of this type are called plate-and-frame; when used in open loops, these heat exchangers are normally of the gasketed type to allow periodic disassembly, cleaning, and inspection. There are many types of permanently-bonded plate heat exchangers, such as dip-brazed and vacuum-brazed plate varieties, and they are often specified for closed-loop applications such as refrigeration. Plate heat exchangers also differ in the types of plates that are used, and in the configurations of those plates. Some plates may be stamped with "chevron" or other patterns, where others may have machined fins and/or grooves.

- **Regenerative Heat Exchanger**

[8] A third type of heat exchanger is the regenerative heat exchanger. In this, the heat (heat medium) from a process is used to warm the fluids to be used in the process, and the same type of fluid is used either side of the heat exchanger (these heat exchangers can be either plate-and-frame or shell-and-tube construction). These exchangers are used only for gases and not for liquids. The major factor for this is the heat capacity of the heat transfer matrix.

- **Plate Fin Heat Exchanger**

[9] This type of heat exchanger uses "sandwiched" passages containing fins to increase the effectivity of the unit. The designs include crossflow and counterflow coupled with various fin configurations such as straight fins, offset fins and wavy fins.

[10] Plate and fin heat exchangers are usually made of aluminium alloys which provide higher heat transfer efficiency. The material enables the system to operate at a lower temperature and reduce the weight of the equipment. Plate and fin heat exchangers are mostly used for low temperature services such as natural gas, helium and oxygen liquefaction

plants, air separation plants and transport industries such as motor and aircraft engines.

Central Heating

[11] A central heating system provides warmth to the whole interior of a building (or portion of a building) from one point to multiple rooms. When combined with other systems in order to control the building climate, the whole system may be a HVAC system.

[12] Central heating differs from local heating in that the heat generation occurs in one place, such as a furnace room in a house or a mechanical room in a large building (though not necessarily at the "central" geometric point). The most common method of heat generation involves the combustion of fossil fuel in a furnace or boiler. The resultant heat then gets distributed: typically by forced-air through ductwork, by water circulating through pipes, or by steam fed through pipes. Increasingly, buildings utilize solar-powered heat sources, in which case the distribution system normally uses water circulation.

Hot-Water Heating

[13] Water is especially favoured for central-heating systems because its high density allows it to hold more heat and because its temperature can be regulated more easily. A hot-water heating system consists of the boiler and a system of pipes connected to radiators, piping, or other heat emitters located in rooms to be heated. The pipes, usually of steel or copper, feed hot water to radiators or convectors, which give up their heat to the room. The water, now cooled, is then returned to the boiler for reheating.

[14] **Two important requirements of a hot-water system are: (1) provision to allow for the expansion of the water in the system, which fills the boiler, heat emitters, and piping, and (2) means for allowing air to escape by a manually or automatically operated valve.** Early hot-water systems, like warm-air systems, operated by gravity, the cool water, being more dense, dropping back to the boiler, and forcing the heated lighter water to rise to the radiators. Neither the gravity warm-air nor gravity hot-water system could be used to heat rooms below the furnace or boiler. Consequently, motor-driven pumps are now used to drive hot water through the pipes, making it possible to locate the boiler at any elevation in relation to the heat emitters. As with warm air, smaller pipes can be used when the fluid is pumped than with gravity operation.

[15] A sealed system provides a form of central heating in which the water used for heating usually circulates independently of the building's normal water supply. A pressure vessel contains compressed gas, separated from the sealed-system water by a diaphragm. This allows for normal variations of pressure in the system. A safety valve allows water to escape from the system when pressure becomes too high, and a valve can open to replenish water from the normal water supply if the pressure drops too low. Sealed systems offer an alternative to open-vent systems, in which steam can escape from the system, and gets replaced from the building's water supply via a feed and central storage system.

Steam Heating

[16] Steam systems are those in which steam is generated, usually at less than 35 kPa

in the boiler, and the steam is led to the radiators through steel or copper pipes. The steam gives up its heat to the radiator and the radiator to the room, and the cooling of the steam condenses it to water. The condensate is returned to the boiler either by gravity or by a pump. The air valve on each radiator is necessary to allow air to escape; otherwise it would prevent steam from entering the radiator. In this system, both the steam supply and the condensate return are conveyed by the same pipe. More sophisticated systems use a two-pipe distribution system, keeping the steam supply and the condensate return as two separate streams.

[17] **Steam's chief advantage, its high heat-carrying capacity, is also the source of its disadvantages. The high temperature (about 102℃) of the steam inside the system makes it hard to control and requires frequent adjustments in its rate of input to the rooms.** To perform most efficiently, steam systems require more apparatus than do hot-water or warm-air systems, and the radiators used are bulky and unattractive. As a result, warm air and hot water have generally replaced steam in the heating of homes built from the 1930s and '40s.

[18] Hydronic heating systems are systems that circulate a medium for heating. Hydronic radiant floor heating systems use a boiler or district heating to heat up hot water and a pump to circulate the hot water in plastic pipes installed in a concrete slab. The pipes, embedded in the floor, carry heated water that conducts warmth to the surface of the floor where it broadcasts energy to the room.

[19] Hydronic systems circulate hot water for heating. Steam heating systems are similar to heating water systems, except steam is used as the heating medium instead of water.

[20] Hydronic heating systems generally consist of a boiler or district heating heat exchanger, hot water circulating pumps, distribution piping, and a fan coil unit or a radiator located in the room or space. Steam heating systems are similar except no circulating pumps are required.

[21] Hydronic systems are closed loop: the same fluid is heated and then reheated. Hydronic heating systems are also used with antifreeze solutions in ice and snow melt systems for walkways, parking lots and streets. They are more commonly used in commercial and whole house radiant floor heat projects, while electric radiant heat systems are more commonly used in smaller "spot warming" applications.

Electric and Gas-Fired Heaters

[22] Electric heating or resistance heating converts electricity directly to heat. Electric heat is often more expensive than heat produced by combustion appliances like natural gas, propane, and oil. Electric resistance heat can be provided by baseboard heaters, space heaters, radiant heaters, furnaces, wall heaters, or thermal storage systems.

[23] Electric heaters are usually part of a fan coil which is part of a central air conditioner. They circulate heat by blowing air across the heating element which is supplied to the furnace through return air ducts. Blowers in electric furnaces move air over one to five

resistance coils or elements which are usually rated at five kilowatts. The heating elements activate one at a time to avoid overloading the electrical system. Overheating is prevented by a safety switch called a limit controller or limit switch. This limit controller may shut the furnace off if the blower fails or if something is blocking the air flow. The heated air is then sent back through the home through supply ducts.

District Heating

[24] **District heating is a system for distributing heat generated in a centralized location for residential and commercial heating requirements such as space heating and water heating. The heat is often obtained from a cogeneration plant burning fossil fuels but increasingly biomass, although heat-only boiler stations, geothermal heating and central solar heating are also used, as well as nuclear power.** District heating plants can provide higher efficiencies and better pollution control than localized boilers. According to some research, District Heating with Combined Heat and Power - CHPDH is the cheapest method of cutting carbon, and has one of the lowest carbon footprints of all fossil generation plants (Fig. 4).

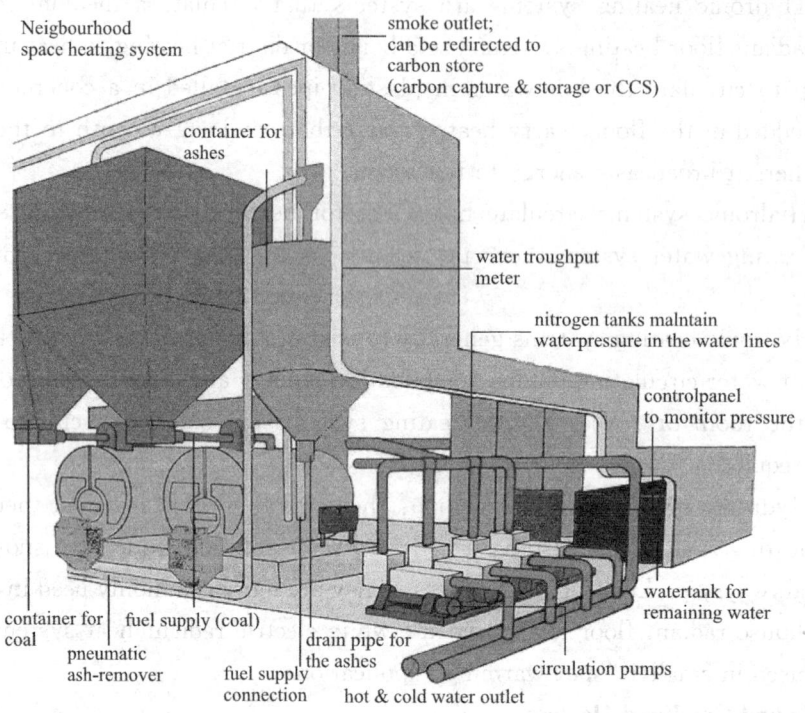

Fig. 4　A district heating system

Heat Generation

[25] The core element of a district heating system is usually a cogeneration plant (also called combined heat and power, CHP) or a heat-only boiler station. Both have in common that they are typically based on combustion of primary energy carriers. The difference between the two systems is that, in a cogeneration plant, heat and electricity are generated simultaneously, whereas in heat-only boiler stations - as the name suggests - only heat is generated.

[26] The combination of cogeneration and district heating is very energy efficient. A thermal power station which generates only electricity can convert less than approximately 50% of the fuel input into electricity. The major part of the energy is wasted in form of heat and dissipated to the environment. A cogeneration plant recovers that heat and can reach total energy efficiency beyond 90%. Other heat sources for district heating systems can be geothermal heat, solar heat, surplus heat from industrial processes, and nuclear power.

[27] A cancelled Russian nuclear district heating plant in Fedyakovo, Nizhny Novgorod Oblast. Nuclear energy can be used for district heating. The principles for a conventional combination of cogeneration and district heating applies the same for nuclear as it does for a thermal power station. One use of nuclear heat generation was with the Agesta Nuclear Power Plant in Sweden. In Switzerland, the Beznau Nuclear Power Plant provides heat to about 20,000 people. Russia has several cogeneration nuclear plants which together provided 456 kW of district heat in 2005. Russian nuclear district heating is planned to nearly triple within a decade as new plants are built.

Heat Distribution

[28] After generation, the heat is distributed to the customer via a network of insulated pipes. District heating systems consists of feed and return lines. Usually the pipes are installed underground but there are also systems with overground pipes. Within the system heat storages may be installed to even out peak load demands.

[29] The common medium used for heat distribution is water, but also steam is used. The advantage of steam is that in addition to heating purposes it can be used in industrial processes due to its higher temperature. The disadvantage of steam is a higher heat loss due to the high temperature. Also, the thermal efficiency of cogeneration plants is significantly lower if the cooling medium is high temperature steam, causing smaller electric power generation. Heat transfer oils are generally not used for district heating, although they have higher heat capacities than water, as they are expensive, and have environmental issues.

[30] **At customer level the heat network is connected to the central heating of the dwellings by heat exchangers (heat substations). The water (or the steam) used in the district heating system is not mixed with the water of the central heating system of the dwelling.** Typical annual loss of thermal energy through distribution is around 10%, as seen in Norway's district heating network.

II. Words and Expressions

fireplace	n.	壁炉
central heating		集中供热
apart from		除…之外
petrochemical	a.	石油化学的
countercurrent		逆流

cross-flow		横向流动
perpendicular	a.	垂直的，正交的
corrugation	n.	起皱，皱状
turbulence	n.	湍流
log mean temperature difference		对数平均温度
NTU		传质单元数
tube	n.	管子，导管
stacked	a.	堆叠的，堆叠式
gasket	n.	垫圈，衬垫
periodic	a.	周期的，定期的
chevron	n.	V形臂章
matrix	n.	矩阵，模型
sandwiched	a.	夹在中间的
offset	v.	抵消，弥补，平衡
aluminium alloys		铝合金
helium	n.	氦
liquefaction plant		液化装置
furnace	n.	火炉，熔炉
copper	n.	铜
vessel	n.	容器，器皿
diaphragm	n.	薄膜，隔膜
replenish	v.	补充
apparatu	n.	装置，设备
hydronic	a.	液体循环的
concrete slab		混凝土板
fan coil		风机盘管
antifreeze	n.	防冻剂
propane	n.	丙烷
district heating		区域供暖
cogeneration	n.	废热发电，热电联产
surplus	a.	剩余的，过剩的

III. Notations

1. Such a source, along with related methods such as fireplaces, cast-iron stoves, and modern space heaters fueled by gas or electricity, is known as direct heating because the conversion of energy into heat takes place at the site to be heated.

Such a source 指的是前一句中的 an open fire

与壁炉、铸铁炉和现代的燃气或电加热器一样，这样的热源被称为直接采暖，因为能量转化为热量就发生在被加热的地点。

2. Another type of heat exchanger is the plate heat exchanger. One is composed of multiple, thin, slightly-separated plates that have very large surface areas and fluid flow passages for heat transfer. This stacked-plate arrangement can be more effective, in a given space, than the shell and tube heat exchanger.

另一种类型的热交换器是平板式换热器。它由许多薄的分离的平板组成，有较大的表面积和流体通道用于热交换。堆叠式平板换热器在给定的空间比壳管式换热器效果更好。

3. Two important requirements of a hot-water system are: (1) provision to allow for the expansion of the water in the system, which fills the boiler, heat emitters, and piping, and (2) means for allowing air to escape by a manually or automatically operated valve.

mean for   预定做某种用途

热水系统需要满足以下两点：（1）系统中流经锅炉、散热器和管道的水要满足其膨胀的要求；（2）要通过手动或自动操作阀以放掉系统中的空气。

4. Steam's chief advantage, its high heat-carrying capacity, is also the source of its disadvantages. The high temperature (about 102℃) of the steam inside the system makes it hard to control and requires frequent adjustments in its rate of input to the rooms.

蒸汽系统的主要优点是高热容量，但也有其缺点。系统中蒸汽的高温（大约102℃）使其很难被控制，而且需要频繁调节进入房间的热量。

5. District heating is a system for distributing heat generated in a centralized location for residential and commercial heating requirements such as space heating and water heating. The heat is often obtained from a cogeneration plant burning fossil fuels but increasingly biomass, although heat-only boiler stations, geothermal heating and central solar heating are also used, as well as nuclear power.

区域供暖，就像供热和供热水一样，是一个满足住宅和商业供暖需求的系统。尽管只产热的锅炉站、地热和集中太阳能采暖以及核能也使用，但区域供暖的热量常常由燃烧化石燃料及日益增多的生物质联合工厂获得。

6. At customer level the heat network is connected to the central heating of the dwellings by heat exchangers (heat substations). The water (or the steam) used in the district heating system is not mixed with the water of the central heating system of the dwelling.

用户端的热网通过热交换器（换热站）连接到居住地的集中供暖系统。区域供暖系统中使用的热水（或蒸汽）没有与居住地的集中供暖系统的水混合。

IV. Exercises

1. Translate the following sentences into Chinese.

(1) A heat exchanger is a device built for efficient heat transfer from one medium to another. The medium may be separated by a solid wall, so that they never mix, or they may be in direct contact. They are widely used in space heating, refrigeration, air conditioning, power plants, chemical plants, petrochemical plants, petroleum refineries, and natural gas processing.

(2) Plate and fin heat exchangers are mostly used for low temperature services such as natural gas, helium and oxygen liquefaction plants, air separation plants and transport industries such as motor and aircraft engines.

(3) A sealed system provides a form of central heating in which the water used for heating usually circulates independently of the building's normal water supply. A pressure vessel contains compressed gas, separated from the sealed-system water by a diaphragm.

(4) Hydronic heating systems generally consist of a boiler or district heating heat exchanger, hot water circulating pumps, distribution piping, and a fan coil unit or a radiator located in the room or space. Steam heating systems are similar except no circulating pumps are required.

(5) The combination of cogeneration and district heating is very energy efficient. A thermal power station which generates only electricity can convert less than approximately 50% of the fuel input into electricity. The major part of the energy is wasted in form of heat and dissipated to the environment.

(6) Also, the thermal efficiency of cogeneration plants is significantly lower if the cooling medium is high temperature steam, causing smaller electric power generation. Heat transfer oils are generally not used for district heating, although they have higher heat capacities than water, as they are expensive, and have environmental issues.

2. Translate the following sentences into English.

（1）以对流换热为主要方式的供热，称为对流供暖。系统中的散热设备是散热器，因此这种系统也称为散热器供暖系统，它是以对流方式向室内供暖。辐射供暖是以辐射传热为主的一种供暖方式。辐射供暖系统的散热设备主要是金属辐射板或以建筑物部分顶棚、地板或墙壁作为辐射散热面。

（2）按系统循环动力的不同，可分为重力循环系统和机械循环系统。靠水的密度差进行循环的系统，称为重力循环系统；靠机械力进行循环的系统，称为机械循环系统。重力循环热水供暖系统的循环作用压力的大小，取决于水温（水的密度）在循环环路的变化状况。

（3）影响散热器传热系数的因素很多：散热器的制造情况（如采用的材料、几何尺寸、结构形式、表面喷涂等因素）和散热器的使用条件（如使用的热媒、温度、流量、室内客气温度及流速、安装方式及组合片数等因素），都综合地影响散热器的散热性能。

Lesson 6　Gas Supply（燃气供应）

I. Text

[1] Fuel gas can refer to any of several gases burned to produce thermal energy. Natural gas is the most common fuel gas, but others include: town gas, syngas, mond gas, propane, butane, regasified liquified petroleum gas, wood gas, producer gas, water gas. Uncompressed hydrogen or compressed hydrogen may be used as a fuel gas.

[2] Natural gas, as a cleaner burning source of fossil fuel than oil or coal, is now commonly believed to offer part of the solution to climate change and to problems associated with poor air quality. Once considered largely a waste product of oil production, natural gas is currently experiencing a huge increase in demand around the world. **As a plentiful, economically viable and less polluting fuel, natural gas makes sense for developing economies looking for new sources for power generation. There is an abundance of natural gas in the world, but it is a non-renewable resource, the formation of which takes thousands and possibly millions of years.** Therefore, as the use of this fossil fuel is increasing, it is important to understand the availability of its supply.

Formation of Gas

[3] Natural gas comes from organic remains of ancient plants and animals. The erosion process carried these biological remains down rivers and streams onto shorelines, where they are deposited along with mud and silt.

[4] Overtime, sedimentary materials such as mud and sand cover the organic remains. Eventually the organic materials transform into petroleum products due to the intense pressure and heat present in the rock formation.

[5] The oil and gas which migrate through the pores in the sedimentary rock eventually reach an impermeable layer of dense rock and collect in large deposits.

[6] Natural gas is a vital component of the world's supply of energy because of its economic and ecological advantages. When used to generate electricity in modern Combine Cycle Gas Turbines (CCGT) efficiency of $50\% \sim 55\%$ can be achieved compared to $35\% \sim 40\%$ commercial fuel oil steam plant. This translate to cost saving to the consumers of electricity in the industrial, commercial and residential sectors. Natural gas is also one of the cleanest and safest energy sources. Unlike other fossil fuels, burning of natural gas emits almost no sulphur into the atmosphere and it is also a safe source of energy when transported, stored and used.

Components of the Gas Distribution Networks

[7] Generally speaking, gas distribution networks are composed of pipes intended for the transport of natural gas, production, storage and distribution centers, compression

stations, and many other devices like valves and regulators.

Gas Pipelines

[8] The high-efficiency movement of natural gas from producing regions to consumption regions requires an extensive and elaborated transportation system. In many instances, the natural gas produced from a particular well will have to travel a long distance (several hundred kilometers, or even of the thousands) to reach its point of use. The transportation system for natural gas consists of a complex network of pipelines, designed to quick and efficient transportation of natural gas from its origin to areas of high demand for natural gas.

[9] There are essentially three major types of pipelines along the transportation route: the gathering system, the interprovincial pipeline, and the distribution system.

[10] The gathering system consists of low-pressure, short-diameter pipelines that transport raw natural gas from the wellhead to the processing plant. If extracted natural gas from a particular well have high sulfur and carbon dioxide contents (sour gas), a specialized sour gas gathering pipe must be installed. Sour gas is extremely corrosive and dangerous, thus, its transportation from the wellhead to the sweetening plant must be done carefully.

[11] The distribution is the final step in delivering natural gas to end users. Some large industrial, commercial, and power generation customers receive natural gas directly from high capacity interprovincial pipelines and pipelines inside the province (usually by contract with the natural gas marketing companies); most other users receive natural gas from a local distribution center (LDC).

[12] LDCs are companies involved in the delivery of natural gas to consumers within a specific geographic area. LDCs companies typically transport natural gas from delivery points along interprovincial pipelines and pipelines inside the province through thousands of miles of short-diameter distribution pipe. Delivery points to LDCs, especially for large municipal areas, are often termed city gates, and are important market centers for the pricing of natural gas. Typically, LDCs take ownership of the natural gas at the city gate, and deliver it to each individual customer's location of use by city gate stations (CGS).

[13] **This requires an extensive network of small-diameter distribution pipes; due to the transportation infrastructure required to move natural gas to many diverse customers across a reasonably wide geographic area, distribution costs typically make up the majority of natural gas costs for small volume end users.** While large pipelines can reduce unit costs by transmitting large volumes of natural gas, distribution companies must deliver small volumes to many more different locations.

[14] The operational costs of CGSs are divided into fixed and variable cost components. There is no limitation on warehouse's inventory at the beginning and ending of the planning horizon. The demand of the consumer groups is satisfied based on a special policy at the end of planning horizon for each period; however, all the demands should be covered

with respect to the established government policies.

Compression Stations

[15] When gas flows through the network, it suffers energy and pressure losses not only due to the friction between the gas and the inner walls of gas ducts but due to the heat transfer between gas and the environment. It is known that one of the most interesting issues in a gas distribution system is the search of the pressure that should be supplied to the gas from the compression stations to reach the consumption points with the required conditions. If the gas has to be supplied to the consumption points with a specified pressure, the undesired pressure drops along the network must be periodically restored. This task is performed by compression stations (CSs) installed on the network.

[16] The compresstion stations (CSs) are used to provide the gas enough energy to be moved along long distances via a pipeline. A number of turbocompressors (TCs) located in parallel are the principal equipments of the CS. A part of the gas crossing through the station is used as fuel gas for the TCs, but this usually only consumes about 3% or 5% of the total gas transported. The compression stations are responsible for making two important decisions: to increase or to decrease compression in the pipelines, and to start up or to shut down the TC units. Incorrect decisions may increase energy cost or lead to customers' dissatisfaction.

Pressure Regulators

[17] On the other hand, if an increase of gas pressure takes place, certain safety limits could be exceeded. In these cases it is necessary to activate an emergency mechanism to avoid such contingency. To cope with it the pressure regulators installed in the network are used to reduce the pressure to the acceptable values which are within these limits. Just as well as the case in the compressor stations, these devices consume a fraction of the total gas transported by the network.

The Planning of the Natural Gas Industry

[18] As known, the natural gas industry involves three main processes:
• Production, which is mainly concerned with development and exploration of gas reserves. Natural gas is supplied to refineries from different sources, including oil and gas wells and importation from other countries. Fuel and gas are the two main consumable substances obtained in refineries. Some of the remaining gas is moved back to injection wells and the rest is transported to compressor stations. These stations, as the heart of the network, consume natural gas energy to increase gas pressure in the network.
• Transmission, which involves moving a large volume of gas at high pressure over long distances from a gas source to distribution centers.
• Distribution, which means routing gas to individual consumers.

[19] **Both transmission and distribution in the natural gas network are included in the distribution process of this chain. Therefore, raw materials and components are moved from suppliers to manufactures, whereas finished products are moved from the manufacturers to the**

end consumers. However, the natural gas that reaches its destination is not always needed right away and like most other commodities, can be stored for an indefinite period of time. Therefore, it is injected into the underground storage facilities. These storage facilities can be located near consumption centers like local gas companies. On the required time i. e. when shortage is occurring, the injected gas which acts as storage is sent to the related pipelines.

[20] The planning of the industry refers to optimizing the whole network of production, distribution, and resources storage to respond properly to the external conditions, such as orders and demand forecasts, over a short or long time horizon. The goal of the planning in the network is the minimization of costs, which are directly or indirectly affected by the process as follows: extraction costs in wells, operation costs in refineries, underutilization costs in wells and refineries, the operational costs in compressor stations and CGSs, gas transmission costs between entities of the network, holding costs in warehouses, and just in-time delivery costs for the consumers.

[21] An optimal gas supply strategy can only be defined after considering a wide range of purchase, transportation, and storage options over a multi-year time horizon. Then, an appropriate distribution network design can make it easy to achieve a variety of supply chain objectives, such as low cost and high responsiveness. Thus companies in the same industry often select very different distribution network designs. **The gas distribution network starts from refineries, as producers, and ends with different types of consumers. Due to the long distances between these two levels, there are retailers and wholesalers acting as centric levels; therefore, the number of distributors depends on their distances from each other.**

[22] This whole process determines how products are retrieved and transported from the distribution centers (DCs) to the customers and affects both the supply chain costs and the customer's perceptions. Therefore, the planning which decreases costs and increases satisfaction is essential for the industry.

Gas Pipelines' Corrosion Defects

[23] As known, gas transmission pipelines usually have a good safety record. This is due to a combination of good design, materials and operating practices; however, like any engineering structure, pipelines do occasionally fail. The most common causes of damage and failures in gas transmission pipelines are external interference (mechanical damage) and corrosion.

[24] Defects occurring during the fabrication of a pipeline are usually assessed against recognised and proven quality control (workmanship) limits. Pipeline failures are usually related to a breakdown in a system, e.g. the corrosion protection system has become faulty, and a combination of ageing coating, aggressive environment, and rapid corrosion growth may lead to a corrosion failure. This type of failure is not simply a corrosion failure, but a corrosion control system failure. Similar observations can be drawn for failures due to external interference, stress corrosion cracking, etc.

[25] Corrosion is an electrochemical process. It is a time dependent mechanism and depends on the local environment within or adjacent to the pipeline. Corrosion usually appears as either general corrosion or localised (pitting) corrosion.

[26] There are many different types of corrosion, including galvanic corrosion, microbiologically induced corrosion. Corrosion causes metal loss. It can occur on the internal or external surfaces of the pipe, in the base material, the seam weld, the girth weld, and/or the associated heat affected zone (HAZ). Internal and external corrosion are together one of the major causes of pipeline failures.

[27] Corrosion in a pipeline may be difficult to characterise. Typically, it will have an irregular depth profile and extend in irregular pattern in both longitudinal and circumferential directions. It may occur as a single defect or as a cluster of adjacent defects separated by full thickness (uncorroded) material. There are no clear definitions of different types of corrosion defects. The simplest and perhaps most widely recognised definitions are as follows:

• pitting corrosion, defined as corrosion with a length and width less than or equal to three times the uncorroded wall thickness.

• general corrosion, defined as corrosion with a length and width greater than three times the uncorroded wall thickness.

Protective Measures

[28] Based on the analysis above, it is becoming increasingly important to ensure safe and failure-free operation of these pipelines. While pipelines are inherently one of the safest mode of transporting bulk energy, with failure rates much less compared to rail/road transportation, failures do occur and sometimes with catastrophic consequences. If the natural gas is accidentally released and ignited, the hazard distance associated with these pipelines to people and property has been found to range from under 20 m for a smaller pipeline at lower pressure, up to over 300 m for a larger one at higher pressure.

[29] The individual risk is defined as the probability of death per year of exposure to an individual at a certain distance from the hazard source. It is usually expressed in the form of iso-risk contours around the hazard source. In the case of pipelines, the iso-risk contour is usually parallel with the pipeline.

[30] Many countries employ numerical criteria in determining acceptability in terms of safety. One approach is to set an upper limit of acceptability and a lower limit of negligible risk; in between is a grey area where risk reduction measures must be considered and discussed on the grounds of reasonableness and cost-benefits. These risk criteria must be consistent with the minimum proximity of the pipeline to normally occupied buildings.

[31] Traditionally, most pipeline operators ensure that during designing process safety provisions are created to provide a theoretical minimum failure rate for the life of the pipeline. Safety provisions govern selection of pipes and other fittings. To prevent corrosion, high-resistance external coating materials electrically isolate a pipeline. **As a secondary protective measure, a low-voltage direct current is impressed in the pipe at precalculated**

distances to transfer any corrosion that occurs as a result of breaks in the coating from buried iron junk, rails, etc. This is called impresses-current cathodic protection (CP).

[32] The quality of the commodity being transported through the line is also ensured. Sometimes, corrosion-preventing chemicals (corrosion inhibitors) are mixed with the commodity.

[33] Regular patrolling of the right of way (ROW) from the air as well as on foot helps prevent deliberate damage of the pipeline in isolated locations. All third-party activities near the route are monitored. Various techniques are routinely used to monitor the status of a pipeline.

[34] Any deterioration in the line may cause a leak or rupture. It is useful to determine the route of a new pipeline at the early planning stages. When a pipeline passes close to a town or any other densely populated area, social risk must be evaluated with individual risk to determine the risk acceptability. Modern methodologies can ensure the structural integrity of an operating pipeline without being taken out of service. **Hence, to manage pipelines effectively, pipeline operators design their pipeline system in a very reliable way by choosing optimal pipeline route and planning pipeline projects efficiently with the consideration of overall profitability of project in long run.** The shortest distance between demand and supply points governs the current practice of feasibility analysis of gas pipeline projects.

[35] A few alternative projects are then developed with the consideration of technical parameters such as throughputs, diameter of pipelines and number of intermediate stations. Subsequent financial analysis selects the best project on the basis of minimum cost with respect to capital and operating cost. The present feasibility analysis never considers maintenance and the augmentation possibility of the pipeline. Both of these factors are highly dependent on the pipeline route.

[36] The optimal gas pipeline route selection is governed by the following goals:

• Establish the shortest possible route connecting the originating, intermediate and terminal locations.

• Ensure, as far as practicable, accessibility during operation and maintenance stages.

• Preserve the ecological balance and avoid/minimize environmental damage. The route should be kept clear of forests as much as possible.

• Avoid densely populated areas for public safety reasons.

• Keep rail, road, river and canal crossings to the bare minimum.

• Avoid hilly or rocky terrain.

• Avoid a route running parallel to high-voltage transmission lines.

• Avoid other obstacles such as wells, houses, orchards, lakes or ponds.

[37] These characteristics must be determined by a reconnaissance survey and the goal of finding the shortest possible route is always important.

II. Words and Expressions

syngas	n.	合成气
mond gas		蒙德煤气，半水煤气
propane	n.	丙烷
butane	n.	丁烷
petroleum	n.	石油
wood gas		木煤气，木瓦斯
makes sense		有意义，言之有理
renewable	a.	可再生的
organic	a.	器官的，有机的
erosion	n.	腐蚀，侵蚀
sedimentary	a.	沉淀的
migrate	v.	移动，移往…
impermeable	a.	不渗透性的，不能渗透的
wellhead	n.	泉源，水源
interprovincial	a.	省与省之间的，省际的
periodically	ad.	周期性的，定期的
contingency	n.	意外事故，偶然性
cope	v.	处理，应付
destination	n.	目的地，终点
optimize	v.	优化，完善
underutilization	n.	未充分利用
responsiveness	n.	响应性，反应性
interference	n.	干扰，冲突
electrochemical	a.	电化学的
seam weld		滚焊，缝熔接
girth weld		环形焊缝
circumferential	a.	周围的
inherently	ad.	固有地，天生地，内在地
catastrophic	a.	灾难的，悲惨的
cathodic	a.	阴极的，负极的
deterioration	n.	恶化，退化
rupture	n.	破裂，决裂
reconnaissance	n.	侦查，勘测

III. Notations

1. As a plentiful, economically viable and less polluting fuel, natural gas makes sense for developing economics looking for new sources for power generation. There is an abun-

dance of natural gas in the world, but it is a non-renewable resource, the formation of which takes thousands and possibly millions of years.

作为一种丰富的、经济上可行的和低污染的燃料，天然气对于发展经济和寻求一种新资源发电来说都很有意义。世界上天然气的含量很丰富，但是它不是可再生资源，它的形成需要几千年、也可能百万年的时间。

2. This requires an extensive network of small-diameter distribution pipes; due to the transportation infrastructure required to move natural gas to many diverse customers across a reasonably wide geographic area, distribution costs typically make up the majority of natural gas costs for small volume end users.

This 指的是 local distribution center.　　make up 组成

由于把天然气运输到许多不同的用户横跨了相当宽的地域，这需要大量的由小直径分配管道组成的管网，对于少量的终端使用者来说，运输成本占天然气成本的大部分。

3. Both transmission and distribution in the natural gas network are included in the distribution process of this chain. Therefore, raw materials and components are moved from suppliers to manufacturers, whereas finished products are moved from the manufacturers to the end consumers.

this chain 指的是前文中的 three main processes

天然气管网的传输和分配包含于这个链条的分配过程。因此，原材料和部件从供应商运输到制造商，而成品则从制造商被分配到终端用户。

4. The gas distribution network starts from refineries, as producers, and ends with different types of consumers. Due to the long distances between these two levels, there are retailers and wholesalers acting as centric levels; therefore, the number of distributors depends on their distances from each other.

each other 指的是 refineries and consumers

燃气输送管网开始于精炼厂，终止于各种各样不同类型的用户。由于这两者之间的距离很长，所以中间会有很多的零售商和批发商，因此，分销商的数量取决于两者之间的距离。

5. As a secondary protective measure, a low-voltage direct current is impressed in the pipe at precalculated distances to transfer any corrosion that occurs as a result of breaks in the coating from buried iron junk, rails, etc.

as a result of　　作为…的结果　　break in　　打断，闯入

作为二级保护措施，管道中保持低压直流电流很重要，管道的预算距离要用来转移任何形式的腐蚀现象，比如，打断地下的铁质垃圾和铁轨等造成的腐蚀。

6. Hence, to manage pipelines effectively, pipeline operators design their pipeline system in a very reliable way by choosing optimal pipeline route and planning pipeline projects efficiently with the consideration of overall profitability of project in long run.

因此，为了使管道更有效的运行，管道操作者从长远利益考虑，通过选择最佳的管道路线和效率高的管道项目，使用可行的方式设计他们的管道系统。

IV. Exercises

1. Translate the following sentences into Chinese.

(1) Natural gas is also one of the cleanest and safest energy sources. Unlike other fossil fuels, burning of natural gas emits almost no sulphur into the atmosphere and it is also a safe source of energy when transported, stored and used.

(2) Some large industrial, commercial, and power generation customers receive natural gas directly from high capacity interprovincial pipelines and pipelines inside the province (usually by contract with the natural gas marketing companies); most other users receive natural gas from a local distribution center (LDC).

(3) It is known that one of the most interesting issues in a gas distribution system is the search of the pressure that should be supplied to the gas from the compression stations to reach the consumption points with the required conditions.

(4) Corrosion is an electrochemical process. It is a time dependent mechanism and depends on the local environment within or adjacent to the pipeline. Corrosion usually appears as either general corrosion or localised (pitting) corrosion.

(5) Traditionally, most pipeline operators ensure that during designing process safety provisions are created to provide a theoretical minimum failure rate for the life of the pipeline. Safety provisions govern selection of pipes and other fittings.

(6) A few alternative projects are then developed with the consideration of technical parameters such as throughputs, diameter of pipelines and number of intermediate stations. Subsequent financial analysis selects the best project on the basis of minimum cost with respect to capital and operating cost.

2. Translate the following sentences into English.

(1) 天然气一般分为四种：从气井开采出来的气田气或称纯天然气；伴随石油一起开采出来的石油气，也称石油伴生气；含石油轻质馏分的凝析气田气；从井下煤层抽出的煤矿矿井气。

(2) 在进行管道内燃气流动的计算时，必须考虑燃气密度的变化。随着沿程压力的下降，燃气的密度也在减少。只在低压管道中燃气密度的变化可忽略不计。

(3) 要使输配系统具有很高的经济指标，除正确选择管径外，还要选择最合理的管道定线方案及燃气调压室、分配站的数目等。在设计过程中常常会遇到很多种设计方案，因此就必须从中选出一个最佳方案。

Extensive Reading

The Invention of District Heating

[1] The inventor of district heating, Birdsill Holly (1820-1894), an American engineer and self-made man, was involved in a variety of different fields: water supply for fire protection, district steam heating, and the building of a steel-framed skyscraper, first on

Goat Island (Niagara Falls), then on Long Island. His first patent, in 1849, was for a pump, as would be most of his patents during the next fifteen years. He also worked on a steampowered fire engine, which could produce a uniform and steady stream of water for fighting fires. After a major fire in Lockport, in 1854, Birdsill Holly set out the Holly System of direct and permanent pressure water supply, more efficient than the reservoirs it replaced. Two years before the Chicago fire of 1871, this city's department of public works had recommended the adoption of the Holly System, but this initiative was not pursued. At the end of the 1870s, Holly undertook a completely new project, a nineteen- story skyscraper on Goat Island in Niagara Falls, he writes:

[2] The original blueprints revealed a building base of 140 feet square. Elevators, located within artistic columns on each side of the skyscraper, would soar 700 feet above the ground. Winding staircases, parallel top the elevator shafts, would contain 1200 steps to the top. The plans called for 368 rooms to be located on the fifth and sixth floors at the top of the skyscraper there would be a promenade, an observatory, and a little train drawn by a locomotive."

[3] Nevertheless, the owners of Goat Island, the Porters, wanting to preserve the island, prevented the scheme's execution. Holly took his blueprints to New York, where he tried to carry out his project on Long Island, but without any success, being considered as a lunatic "farmer from the West."

[4] After this fruitless experience Birdsill Holly returned to plumbing. In 1876 he tried out his idea of distance heating by steam in his own garden in Lockport. He was able to solve problems such as expansion joints, regulatory systems, steam meters, pressure, protection against condensation, and insulation. The year after, incorporated as the Holly Steam Combination Company, he laid out pipes connected to a boiler and a pump throughout the Lockport town center; as was reported a few years later, this network reached 65 houses with 3 miles of pipes. In 1881, several towns, notably New York City, tried out this new procedure for heating buildings within a given area. The same year, Holly deposited several patents which described the respective features the system: mains pipes, meters and steam-pressure regulator. His interest in steam networks was a logical progression from working on pipes, pumps, and fire fighting systems, but his project for the tower on Goat Island is more surprising and reveals the open-minded character of the inventor.

A Brief History of Radiant Panel Heating

[1] The first recorded use of radiant heating in buildings was by the ancient Romans who used it in their well-known bathhouses. These buildings were designed with chambers that allowed fires to be placed under the floors and within walls. The hot exhaust gases were routed through flue-like chambers, giving up their heat to the masonry surfaces along the way. Patrons were not only bathed in warm water, but the radiant heat from the warm walls and floors as well. As crude as these systems were by today's standards, they obvi-

ously offered comfort unattainable by simpler means. Their intricate construction testifies that comfort was highly valued, and justified the elaborate construction necessary to achieve it.

Hydronic Radiant Pipe in the Early 1900's

[2] The first use of hydronic radiant floor heating using iron pipe was in England around 1907. In the years that followed, the English applied it in all types of buildings, including stores, schools, and hospitals.

[3] Hydronic floor heating gained exposure in the US through the well-known designs of architect Frank Lloyd Wright, among others. Wright understood the need to provide superior comfort as part of the experience of living in a one-of-a-kind home. He used radiant heating to transform the massive stone and masonry surfaces typical of his designs into inviting elements, rather than the harsh cold surfaces they are often perceived as. Concerning radiant heating he wrote: "No heating was visible nor was it felt directly as such. It was really a matter not of heating at all, but an affair of climate."

[4] During the post-war housing boom of the late 1940's, thousand of hydronic floor heating systems were installed in the United States using copper tubing and steel pipe. Although some of these systems are still operating today, many failed due to metal fatigue or chemical incompatibilities with concrete. When a leak occurred, it was often difficult or impossible to find and correct. Eventually many of these early systems were simply abandoned in favor of other methods of heating. Skepticism understandably developed and interest in hydronic floor heating declined. The "death blow" to the first generation of hydronic floor heating was the advent of central air conditioning in the 1960's. Since a forced-air system could transport both heated air in winter and cooled air in summer, and didn't suffer from the potential leakage problems of early floor heating systems, it quickly became the standard of the American housing industry.

[5] Rapid growth in the US home building industry over the last 40 years spawned a very competitive and price-dominated market for heating systems, a market that willingly substituted lower cost, and faster installation, for proven comfort. Many homebuyers were apathetic about the type of heat in their new house. Most just assumed their builder knew the best way to heat their house and accepted his choice without question. Unfortunately, many became "casualties" of an apathetic attitude toward comfort that accompanied the fasttract building ethic. Many would now be more discriminating about the method of heating used in any new home or addition they build.

[6] A recent survey conducted by an independent consumer research firm found that 80% of the 80,000 American households responding were only somewhat or not at all satisfied with their heating system. Only 16% said that price was the most important factor in purchasing a new heating system. The message is clear: Most people want better comfort from their home's heating system, and most would willingly spend more money to achieve it.

[7] Without a doubt, comfort has always been the most sought-after benefit of radiant

panel heating. Even early hydronic radiant heating systems, though sometimes plagued by premature material failures, delivered superior comfort for their time. A way was needed to retain this comfort, while at the same time making hydronic floor heating easy to install, and capable of a long, trouble-free life.

[8] During the 1960's, as American interest in floor heating was rapidly declining, research was underway in Europe on a new polymer material called cross-linked polyethylene, or "PEX" for short. Although originally developed as sheathing for underwater cable, it would become the material that eventually revolutionized the hydronic floor heating market worldwide. Today's hydronic radiant panel heating systems are as different from their predecessors as compact disks are from phonograph records. Several kinds of high-tech piping materials now make it possible to install systems that can last for decades, perhaps even longer than the buildings they are a part of. These materials have been extensively tested in radiant panel heating applications over the last three decades. Each year, hundreds of millions of feet of tubing are installed in new radiant panel heating systems worldwide. Major advances have also been made in heat sources and controls. This state-of-the-art hardware has effectively merged the superior comfort aspects of radiant heating with market demands for easy installation and reliability.

Lesson 7 Ventilation System (通风系统)

I. Text

[1] Outdoor air that flows through a building is often used to dilute and remove indoor air contaminants. **However, the energy required to condition this outdoor air can be a significant portion of the total space-conditioning load. The magnitude of outdoor airflow into the building must be known for proper sizing of the HVAC equipment and evaluation of energy consumption.** For buildings without mechanical cooling and dehumidification, proper ventilation and infiltration airflows are important for providing comfort for occupants. ASHRAE Standard 55 specifies conditions under which 80% or more of the occupants in a space will find it thermally acceptable.

Introduction of Basic Concepts and Terminology

[2] Air exchange of outdoor air with air already in a building can be divided into two broad classifications: ventilation and infiltration.

[3] Ventilation is intentional introduction of air from the outside into a building; it is further subdivided into natural and mechanical ventilation. Natural ventilation is the flow of air through open windows, doors, grilles, and other planned building envelope penetrations, and it is driven by natural and/or artificially produced pressure differentials. Mechanical (or forced) ventilation, shown in Figure 1, is the intentional movement of air into and out of a building using fans and intake and exhaust vents.

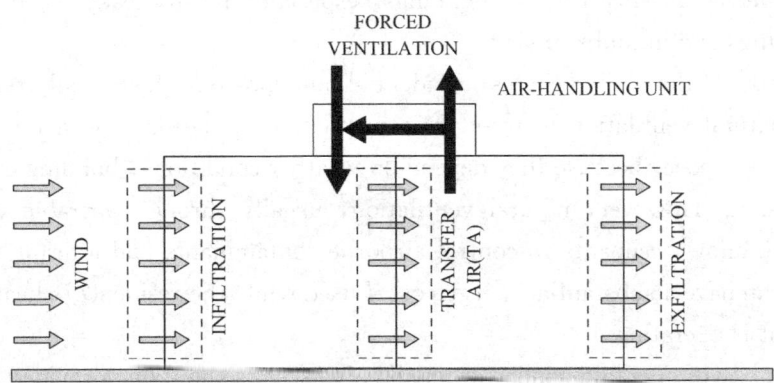

Fig. 1 Two-Space Building with Mechanical Ventilation, Infiltration, and Exfiltration

[4] Infiltration is the flow of outdoor air into a building through cracks and other unintentional openings and through the normal use of exterior doors for entrance and egress. Infiltration is also known as air leakage into a building. Exfiltration, depicted in Figure 1, is leakage of indoor air out of a building through similar types of openings. Like natural

ventilation, infiltration and exfiltration are driven by natural and/or artificial pressure differences. These forces are discussed in detail in the section on Driving Mechanisms for Ventilation and Infiltration. Transfer air is air that moves from one interior space to another, either intentionally or not.

[5] **Ventilation and infiltration differ significantly in how they affect energy consumption, air quality, and thermal comfort, and they can each vary with weather conditions, building operation, and use.** Although one mode may be expected to dominate in a particular building, all must be considered in the proper design and operation of an HVAC system.

Ventilation Air

[6] Ventilation air is air used to provide acceptable indoor air quality. It may be composed of mechanical or natural ventilation, infiltration, suitably treated recirculated air, transfer air, or an appropriate combination, although the allowable means of providing ventilation air varies in standards and guidelines.

[7] Modern commercial and institutional buildings normally have mechanical ventilation and are usually pressurized somewhat to reduce or eliminate infiltration. Mechanical ventilation has the greatest potential for control of air exchange when the system is properly designed, installed, and operated; it should provide acceptable indoor air quality and thermal comfort when ASHRAE Standard 55 and 62.1 requirements are followed.

[8] In commercial and institutional buildings, natural ventilation (e.g., through operable windows) may not be desirable from the point of view of energy conservation and comfort. In commercial and institutional buildings with mechanical cooling and ventilation, an air- or water-side economizer may be preferable to operable windows for taking advantage of cool outdoor conditions when interior cooling is required. Infiltration may be significant in commercial and institutional buildings, especially in tall, leaky, or partially pressurized buildings and in lobby areas.

[9] In most of the United States, residential buildings have historically relied on infiltration and natural ventilation to meet their ventilation air needs. Neither is reliable for ventilation air purposes because they depend on weather conditions, building construction, and maintenance. However, natural ventilation, usually through operable windows, is more likely to allow occupants to control airborne contaminants and interior air temperature, but it can have a substantial energy cost if used while the residence's heating or cooling equipment is operating.

[10] In place of operable windows, small exhaust fans should be provided for localized venting in residential spaces, such as kitchens and bathrooms. Not all local building codes require that the exhaust be vented to the outside. Instead, the code may allow the air to be treated and returned to the space or to be discharged to an attic space. Poor maintenance of these treatment devices can make non-ducted vents ineffective for ventilation purposes. Condensation in attics should be avoided. In northern Europe and in Canada, some building codes require general mechanical ventilation in residences, and heat recovery heat exchang-

ers are popular for reducing energy consumption. Low-rise residential buildings with low rates of infiltration and natural ventilation, including most new buildings, require mechanical ventilation at rates given in ASHRAE Standard 62.2.

Forced-Air Distribution Systems

[11] Figure 2 shows a simple air-handling unit (AHU) or air handler that conditions air for a building. Air brought back to the air handler from the conditioned space is return air (RA). The return air either is discharged to the environment [exhaust air (EA)] or is reused [recirculated air (CA)]. Air brought in intentionally from the environment is outdoor or outside air (OA). Because outdoor air may need treatment to be acceptable for use in a building, it should not be called "fresh air." Outside and recirculated air are combined to form mixed air (MA), which is then conditioned and delivered to the thermal zone as supply air (SA). Any portion of the mixed air that intentionally or unintentionally circumvents conditioning is bypass air (BA). Because of the wide variety of air handling systems, the airflows shown in Figure 2 may not all be present in a particular system as defined here. Also, more complex systems may have additional airflows.

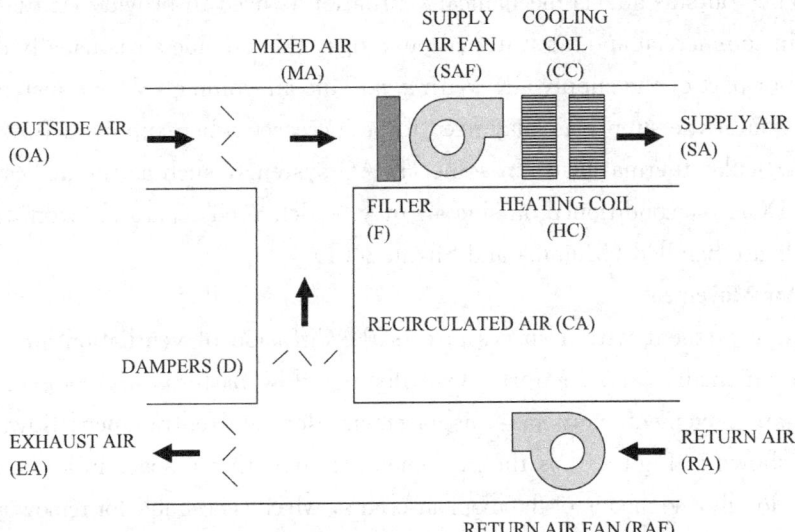

Fig. 2 Simple All-Air Air-Handling Unit with Associated Airflows

Outside Air Fraction

[12] The outside airflow introduced to a building or zone by an air-handling unit can also be described by the outside air fraction X_{oa}, which is the ratio of the volumetric flow rate of outside air brought in by the air handler to the total supply airflow rate:

$$X_{oa} = \frac{Q_{oa}}{Q_{sa}} = \frac{Q_{oa}}{Q_{ma}} = \frac{Q_{oa}}{Q_{oa}+Q_{ca}} \tag{1}$$

where

X_{oa} = the outside air fraction

Q_{oa} = the outside air rate

Q_{sa} = the total supply airflow rate
Q_{ma} = the mixed air rate
Q_{ca} = the recirculated air rate

[13] When expressed as a percentage, the outside air fraction is called the percent outside air. The design outside airflow rate for a building's or zone's ventilation system is found by applying the requirements of ASHRAE Standard 62.1 to that specific building. The supply airflow rate is that required to meet the thermal load. The outside air fraction and percent outside air then describe the degree of recirculation, where a low value indicates a high rate of recirculation, and a high value shows little recirculation. Conventional all-air air-handling systems for commercial and institutional buildings have approximately 10% to 40% outside air. 100% outside air means no recirculation of return air through the air-handling system. Instead, all the supply air is treated outside air, also known as makeup air (KA), and all return air is discharged directly to the outside as relief air (LA), via separate or centralized exhaust fans. An air-handling unit that provides 100% outside air to offset air that is exhausted is typically called a makeup air unit (MAU).

[14] When outside air via mechanical ventilation is used to provide ventilation air, as is common in commercial and institutional buildings, this outside air is usually delivered to spaces as all or part of the supply air. With a variable-air-volume (VAV) system, the outside air fraction of the supply air may need to be increased when supply airflow is reduced to meet a particular thermal load. In some HVAC systems, such as the dedicated outside air system (DOAS), conditioned outside air may be delivered separately from the way the spaces' loads are handled (Mumma and Shank 2001).

Room Air Movement

[15] Air movement within spaces affects the diffusion of ventilation air and, therefore, indoor air quality and comfort. Two distinct flow patterns are commonly used to characterize air movement in rooms: displacement flow and entrainment flow. Displacement flow, shown in Figure 3, is the movement of air within a space in a piston or plug-type motion. Ideally, no mixing of the room air occurs, which is desirable for removing pollutants

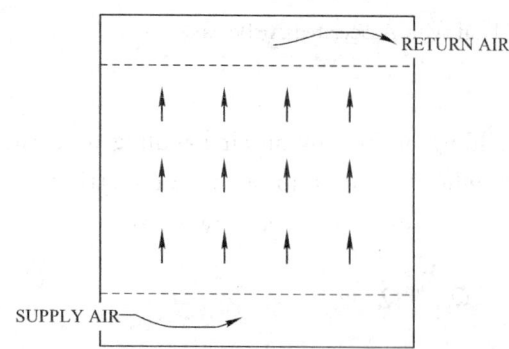

Fig. 3 Displacement Flow within a Space

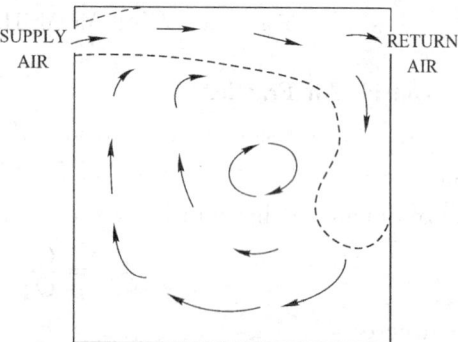

Fig. 4 Entrainment Flow within a Space

generated within a space. A laminar-flow air distribution system that sweeps air across a space may produce displacement flow.

[16] Entrainment flow, shown in Figure 4, is also known as conventional mixing. Systems with ceiling-based supply air diffusers and return air grilles are common examples of air distribution systems that produce entrainment flow. Entrainment flow with very poor mixing in the room has been called *short-circuiting flow* because much of the supply air leaves the room without mixing with room air. There is little evidence that properly designed, installed, and operated air distribution systems exhibit short-circuiting, although poorly designed, installed, or operated systems may short-circuit, especially ceiling-based systems in heating mode (Offer Mann and In-House 1989).

[17] Perfect mixing occurs when supply air is instantly and evenly distributed throughout a space. Perfect mixing is also known as complete or uniform mixing; the air may be called well stirred or well mixed. This theoretical performance is approached by entrainment flow systems that have good mixing and by displacement flow systems that allow too much mixing (Rock et al. 1995).

[18] Under floor air distribution (UFAD or UAD), as shown in Figure 5, is a hybrid method of conditioning and ventilating spaces (Bauman and Daly 2003). Air is introduced through a floor plenum, with or without branch ductwork or terminal units, and delivered to a space by floor-mounted diffusers. These diffusers encourage air mixing near the floor to temper the supply air. The combined air then moves vertically through the space, with reduced mixing, toward returns or exhausts placed in or near the ceiling. This vertical upward movement of the air is in the same direction as the thermal and contaminant plumes created by occupants and common equipment. Ventilation performance for UFAD systems is thus between floor-to-ceiling displacement flow and perfect mixing.

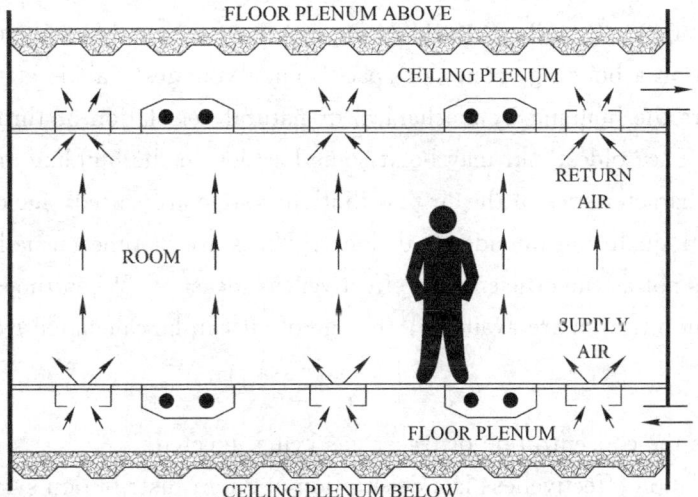

Fig. 5 Underfloor Air Distribution to Occupied Space Above

[19] Supply air that enters a space through a diffuser is also known as primary air. A jet is formed as this primary air leaves the diffuser. Secondary air is the room air entrained into the jet. **Total air is the combination of primary and secondary air at a specific point in a jet. The term *primary air* is also used to describe supply air provided to fan-powered mixing boxes by a central air-handling unit.**

[20] For evaluation of indoor air quality and thermal comfort, rooms are often divided into two portions: the occupied zone and the remaining volume of the space. Often, this remaining volume is solely the space above the occupants and is referred to as the ceiling zone. The occupied zone is usually defined as the lowest 6 ft of a room, although layers near the floor and walls are sometimes deducted from it. Ceiling and floor plenums are not normally included in the occupied or ceiling zones. Thermal zones are different from these room air zones, and are defined for HVAC subsystems and their controls.

Calculation of Correlational Parameter

[21] The air exchange (or change) rate I compare airflow to volume and is

$$I = Q/V \tag{2}$$

where

Q = volumetric flow rate of air into space, cfm

V = interior volume of space, ft^3

The air exchange rate has units of 1/time, usually h^{-1}. When the time unit is hours, the air exchange rate is also called air changes per hour (ach).

[22] Time constantst τ, which have units of time (usually in hours or seconds), are also used to describe ventilation and infiltration. One time constant is the time required for one air change in a building, zone, or space if ideal displacement flow existed. It is the inverse of the air exchange rate:

$$\tau = 1/I = V/Q \tag{3}$$

The age of air θ_{age} (Sandberg 1981) is the length of time t that some quantity of outside air has been in a building, zone, or space. The "youngest" air is at the point where outside air enters the building by mechanical or natural ventilation or through infiltration (Grieve 1989). The "oldest" air may be at some location in the building or in the exhaust air. When the characteristics of the air distribution system are varied, age of air is inversely correlated with quality of outside air delivery. Units are of time, usually in seconds or minutes, so it is not a true efficiency or effectiveness measure. When time-dependent data of tracer gas concentration are available, the age of air can be calculated from:

$$\theta_{age} = \int_{\theta=0}^{\infty} \frac{C_{in} - C}{C_{in} - C_0} d\theta \tag{4}$$

where C_{in} is the concentration of tracer gas being injected.

[23] Ventilation effectiveness is a description of an air distribution system's ability to remove internally generated pollutants from a building, zone, or space. Air change effectiveness is a description of an air distribution system's ability to deliver ventilation air to a

building, zone, or space. Air change effectiveness measures ε_I are no dimensional gages of ventilation air delivery. One common definition of air change effectiveness is the ratio of a time constant to an age of air:

$$\varepsilon_I = \tau/\theta_{age} \tag{5}$$

[24] An HVAC design engineer often assumes that a properly designed, installed, operated, and maintained air distribution system provides an air change effectiveness of about 1. ASHRAE Standard 129 describes a method for measuring air change effectiveness of mechanically vented spaces and buildings with limited air infiltration, exfiltration, and air leakage with surrounding indoor spaces.

II. Words and Expressions

ventilation	n.	通风
contaminant	n.	杂质（污染物质，沾染物）
magnitude	n.	大小，重要，光度，（地震）级数
evaluation	n.	估价，评价
infiltration	n.	浸润，渗透
occupant	n.	占有者，居住者，占领者
exhaust	n.	排气，排气装置
	v.	用尽，耗尽，使…精疲力尽；[计算机] 排除
unintentional	adj.	非故意的，无心的
maintenance	n.	维护，保持，维修，生活费用
airborne	adj.	空运的，空中传播的
interior	n.	内部
	a.	内部的，内地的，国内的，在内的
	adj.	大量的，实质上的，有内容的
localize	vi.	集中，局部化
air-handling unit		空调器，空气调节器
airflow	n.	气流
additional	adj.	附加的，另外的
volumetric	adj.	测定体积的
ASHRAE		American Society of Heating, Refrigerating and Air-Conditioning Engineers 美国采暖制冷与空调工程师协会
recirculation	n.	再循环（回流，信息重记，信息重复循环）
pattern	n.	图案，式样，典范
	v.	仿造，模仿

laminar-flow	n.	层流
distribution	n.	分发,分配,散布,分布
sweep	n.	扫除,席卷,范围
	v.	扫除,用手指弹,掠过
diffuser	n.	扩散器(漫射体,扬声器纸盆,传播者)
entrainment	n.	带去(夺取,雾沫)
theoretical	adj.	理论上的
hybrid	n.	混血儿,杂种,混合物
	adj.	混合的,杂种的,混合语的
non-dimensional	adj.	无量纲化的
maintain	v.	维持,维修,保养,坚持
air change effectiveness		换气效率
air leakage		漏气量
surrounding	adj.	周围的
	n.	环境,周围的事物

III. Notations

1. However, the energy required to condition this outdoor air can be a significant portion of the total space-conditioning load. The magnitude of outdoor airflow into the building must be known for proper sizing of the HVAC equipment and evaluation of energy consumption.

然而,用于处理室外空气所需的能量占空调总负荷的很重要一部分。室外空气的流入量必须已知,为合理的确定HVAC系统设备尺寸及系统能耗评价提供依据。

2. Ventilation and infiltration differ significantly in how they affect energy consumption, air quality, and thermal comfort, and they can each vary with weather conditions, building operation, and use.

通风和渗透的显著的不同在于它们对能耗、空气质量、热舒适性的影响方式不同,并且它们也会因天气状况、建筑物的使用情况等不同而影响不同。

3. Total air is the combination of primary and secondary air at a specific point in a jet. The term *primary air* is also used to describe supply air provided to fan-powered mixing boxes by a central air-handling unit.

总空气量是指在某一状态下一次回风和二次回风的总和。"一次回风"也被用于描述集中空气处理单元提供给风机混合箱提供的那部分送风气流。

IV. Exercises

1. Translate the following sentences into Chinese.

(1) Ventilation is intentional introduction of air from the outside into a building; it is further subdivided into natural and mechanical ventilation. Natural ventilation is the flow of air through open windows, doors, grilles, and other planned building envelope penetra-

tions, and it is driven by natural and/or artificially produced pressure differentials. Mechanical (or forced) ventilation, shown in Figure 1, is the intentional movement of air into and out of a building using fans and intake and exhaust vents.

(2) In most of the United States, residential buildings have historically relied on infiltration and natural ventilation to meet their ventilation air needs. Neither is reliable for ventilation air purposes because they depend on weather conditions, building construction, and maintenance. However, natural ventilation, usually through operable windows, is more likely to allow occupants to control airborne contaminants and interior air temperature, but it can have a substantial energy cost if used while the residence's heating or cooling equipment is operating.

(3) When outside air via mechanical ventilation is used to provide ventilation air, as is common in commercial and institutional buildings, this outside air is usually delivered to spaces as all or part of the supply air. With a variable-air-volume (VAV) system, the outside air fraction of the supply air may need to be increased when supply airflow is reduced to meet a particular thermal load. In some HVAC systems, such as the dedicated outside air system (DOAS), conditioned outside air may be delivered separately from the way the spaces' loads are handled (Mumma and Shank 2001).

(4) Air movement within spaces affects the diffusion of ventilationair and, therefore, indoor air quality and comfort. Two distinct flow patterns are commonly used to characterize air movement in rooms: displacement flow and entrainment flow. Displacement flow, shown in Figure 3, is the movement of air within a space in a pistonor plug-type motion. Ideally, no mixing of the room air occurs, which is desirable for removing pollutants generated within a space. A laminar-flow air distribution system that sweeps air across a space may produce displacement flow.

(5) An HVAC design engineer often assumes that a properly designed, installed, operated, and maintained air distribution system provides an air change effectiveness of about 1. ASHRAE Standard 129 describes a method for measuring air change effectiveness of mechanically vented spaces and buildings with limited air infiltration, exfiltration, and air leakage with surrounding indoor spaces.

2. Translate the following sentences into English.

(1) 保证室内空气品质的主要措施是通风，即用污染物浓度很低的室外空气置换室内污染物浓度较高的空气。其所需的通风量应根据稀释室内污染物达到标准规定的浓度的原则确定。对于以人群活动为主的建筑，主要污染源是人群。因此，这类建筑都是以人来确定必需的通风量——新风量，即用稀释人体散发的二氧化碳量来确定新风量。为了同时考虑稀释人员活动引起的其他污染物和气味，许多国家都把二氧化碳浓度控制在0.1%，世界卫生组织（WHO）建议0.25%。

(2) 空调建筑通常是一个密闭性很好的建筑，如果没有合理的通风，其空气品质还不如通风良好的普通建筑。在空调建筑中，除了工艺过程排放有害气体需专项治理外，一般的通风问题由空调系统来承担。在空气—水系统中，通常设有专门的新风系统，给各个房间送新风，以承担建筑的通风和改善空气品质的任务。全空气系统应引入室外新风，与回风共同处理后下能够入室内，稀释室内污染物。

（3）局部排风是直接从污染源处排除污染物的一种局部通风方式。当污染物集中于某处发生时，局部排风是最有效的治理污染物对环境危害的通风方式。如果这种场合采用全面通风方式，反而使污染物在室内扩散；当污染物发生量大时，所需的稀释通风量则过大，甚至在实际上难于实现。局部排风系统主要由排风罩、风机、空气净化器、风管、排风口组成。

（4）依靠热压或风压为动力的自然通风是人们应用广泛的一种通风方式，一般的居住建筑、普通办公楼、工业厂房等的室内空气品质保证都主要依靠自然通风。然而，自然通风又是难于进行有效控制的通风方式，我们只有通过对自然通风方式的基本原理的了解，采取一定得措施，才能使自然通风基本上按照预想的模式进行。

（5）对房间进行通风，实际上风量总是自动平衡的，我们这里指的"空气平衡"是按设计者或使用者的意愿进行的有计划的平衡。如果不做空气平衡的设计，有可能在实际运行时所达到的平衡状态达不到通风的要求。例如，在一个房间内为排除某污染源散发的污染物而安装一套局部排风系统，但运行时不好用，风量达不到要求，其问题在于该房间在地下室，密闭性较好，由于没有相应的进风系统或进风通道，致使房间负压较大，排风系统风量减小。

Lesson 8 Air-Conditioning System (空调系统)

I. Text

[1] The objective of an air-conditioning system is to provide a comfortable environment for an occupant or occupants of a residential, public, medical, factory, or office buildings. **The comfort environment is the result of simultaneous control of temperature, humidity, cleanliness and air distribution with the occupant's vicinity that includes the proper acoustic level.** Demands for close control of the environment required new approaches for HVAC systems.

Introduction of Basic Concepts and Terminology

[2] HVAC systems contain five kinds of different forms, which are all-air system, air-water system, all-water system and direct expansion system.

[3] All-air system is one in which the air is treated in a central refrigeration plant. The cold air is supplied to a space via ducts and distributed by means of terminal outlets mixing terminal outlets. There are no additional cooling loads at the treated space.

[4] Air-water system, in which the air apparatus and refrigeration plants are separate from the conditioned space. However, the cooling and heating of the conditioned space is affected in only a small part by air brought from the central apparatus. The major portion of the room thermal load is balanced by warm or cool water circulated either through a coil in an induction unit or through a radiant panel.

[5] All-water system is that with fan-coil room terminals to which are connected one or two water circuits. The cooling medium (such as chilled water or brine) may be supplied from a remote source and circulated through the coil of fan coil terminal, which is located in the conditioned space.

[6] Direct expansion system is widely used in a "packaged form", usually sized in small or intermediate tonnage ranges. **Heat is removed by a simple method using a finned tube coil with direct expansion of the refrigerant into the coil, hence, the term Dx. In general system, a Dx system can be described as a refrigeration system where the cooling effect is obtained directly from the refrigerant.**

Introduction of Basic System

[7] A multitude of different systems which vary according to application and are modular in many respects so that adding to subtracting different components can satisfy different design conditions. The essential elements of an air conditioning system are illustrated in Fig 1. The basic elements, optional components and their functions are listed in Table 1.

    O. A. —— outdoor air   B. A. —— bypass air
    R. A. —— return air   S. A. —— supply air

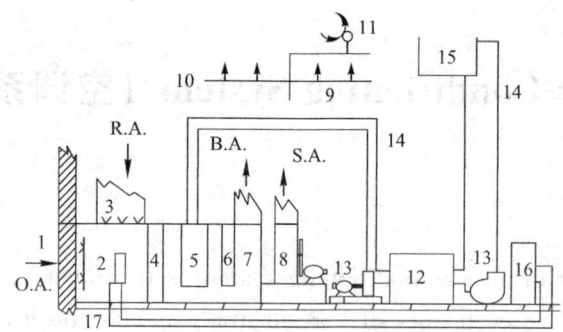

Fig. 1 The essential elements of an air conditioning system

System components   Table 1

System components	Function performed
Air side	
1. Outdoor air intake	Path for outdoor used air for ventilation and marginal weather cooling
2. Preheated	Preheated air
3. Return air intake	Path for return and/or recirculated air to apparatus
4. Filter	Removing contaminants from air
5. Dehumidifier	Cooling and dehumidifying
6. Heating coil	Heating in winter and reheat for temperature and/or humidity control
7. Humidifier	Humidifying
8. Fan	Air propulsion
9. Duct system	Path for air transmission
10. Air outlet	Air distribution within air conditioned space
11. Air terminal	Enclosure for air handling
12. Refrigeration machine	Means for cooling
13. Pump	Water or brine propulsion
14. Water or brine piping	Patch for transmission of water or brine between heat exchangers
15. Cooling tower	Heat disposal from water used in condensing refrigerant
16. Boiler an auxiliaries	Provides steam or hot water
17. Piping	Path for transmission of steam or hot water

Room Air Diffusion

[8] Room air distribution systems are intended to provide thermal comfort and ventilation for space occupants and processes (Fig. 2). Although air terminals (inlets and out-

lets), terminal units, local ducts, and rooms themselves may affect room air diffusion, this part addresses only air terminals and their direct effect on occupant comfort. This part is intended to present HVAC designers the fundamental characteristics of air distribution devices.

[9] Room air diffusion methods can be classified as one of the following:

• Mixed systems produce little or no thermal stratification of air within the space. Overhead air distribution is an example of this type of system.

• Fully (thermally) stratified systems produce little or no mixing of air within the occupied space. Thermal displacement ventilation is an example of this type of system.

• **Partially mixed systems provide some mixing within the occupied and/or process space while creating stratified conditions in the volume above. Most under floor air distribution designs are examples of this type of system.**

• Task/ambient conditioning systems focus on conditioning only a certain portion of the space for thermal comfort and/or process control. Examples of task/ambient systems are personally controlled desk outlets (sometimes referred to as personal ventilation systems) and spot-conditioning systems.

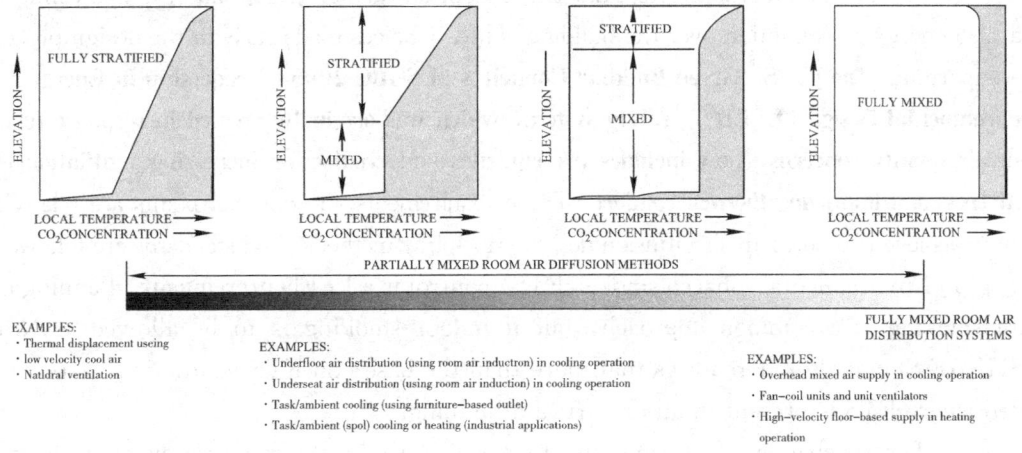

Fig. 2 Classification of Air Diffusion Methods

[10] Air distribution systems, such as displacement ventilation (DV) and underfloor air distribution (UFAD), that deliver air in cooling mode at or near floor level and return air at or near ceiling level produce varying amounts of room air stratification. Figure 1 presents a series of simplified vertical profiles of temperature and pollutant concentration representing the spectrum of stratified conditions that may exist under cooling operation, from fully stratified (e. g. , DV systems) to fully mixed (e. g. , conventional overhead systems). For floor-level supply, thermal plumes that develop over heat sources in the room play a major role in driving overall floor-to-ceiling air motion. The amount of stratification in the room is primarily determined by the balance between total room airflow and heat load. **In practice, the actual temperature (or concentration) profile depends on the combined effects of various factors, but is largely driven by the charactcristics of the room supply airflow and heat load configuration.**

[11] For room supply airflow, the major factors are:
- Total room supply airflow quantity
- Room supply air temperature
- Diffuser type
- Diffuser throws height (or outlet velocity); this is associated with the amount of mixing provided by a floor diffuser (or room conditions near a low-sidewall DV diffuser)

[12] For room heat loads, the major factors are:
- Magnitude and number of loads in space
- Load type (point or distributed source)
- Elevation of load (e. g. , overhead lighting, person standing on floor, floor-to-ceiling glazing)
- Radiative/convective split
- For pollutant concentration profiles, whether pollutants are associated with heat sources

Indoor Air Quality and Sustainability

[13] Air diffusion methods affect not only indoor air quality (IAQ) and thermal comfort, but also energy consumption over the building's life. Choices made early in the design process are important. **The U. S. Green Building Council's (USGBC 2005) Leadership in Energy and Environmental Design (LEED®) rating system, which was originally created in response to indoor air quality concerns, now includes prerequisites and credits for increasing ventilation effectiveness and improving thermal comfort.** These requirements and optional points are relatively easy to achieve if good room air diffusion design principles, methods, and standards are followed.

[14] Environmental tobacco smoke (ETS) control is a LEED prerequisite. Banning indoor smoking is a common approach, but if indoor smoking is to be allowed, ANSI/ASHRAE Standard 62. 1 requires that more than the base non-ETS ventilation air be provided where ETS is present in all or part of a building.

[15] The air change effectiveness is affected directly by the room air distribution system's design, construction, and operation, but is very difficult to predict. Many attempts have been made to quantify air change effectiveness, including ASHRAE Standard 129. However, this standard is only for experimental tests in well-controlled laboratories, and should not be applied directly to real buildings.

[16] ANSI/ASHRAE Standard 62. 1-2007 provides a table of typical values to help predict ventilation effectiveness. **For example, well designed ceiling-based air distribution systems produce near-perfect air mixing in cooling mode, and yield an air change effectiveness of almost 1. 0. Displacement and under floor air distribution (UFAD) systems have the potential for values greater than 1. 0.**

Fully Stratified Systems

[17] Fully stratified air distribution systems have been used in industrial applications for many years. In the 1980s, they became a popular alternative for office and classroom

HVAC in Europe, and their popularity has recently spread to North America because of their high contaminant removal efficiencies and their possible energy savings, especially in relatively mild climates. Thermal displacement ventilation (TDV) systems are the most widely used variant of these systems.

[18] The main objective of a mixed-air system is to create a homogenous mixture of supply and room air throughout the space. Contaminants and heat are diluted and then extracted through the return inlet. TDV systems (Figure 3) do not attempt to mix heat and contaminants; instead, they allow them to escape into the upper uninhabited zone, from which they are extracted. With a TDV system, supply air is introduced directly into the occupied zone at low velocity and a temperature lower than that of room air. Contaminants and heat in the space are carried by convective flows (created by space heat sources) into the upper part of the room. Warm air in the upper zone does not recirculate into the occupied zone, so the temperature and concentration of most impurities at the exhaust inlet exceed those in the occupied zone and at the breathing level. TDV systems offer increased ventilation effectiveness and may reduce HVAC energy consumption. Applications include classrooms, conference rooms, theaters, restaurants, supermarkets, and spaces with high ceilings.

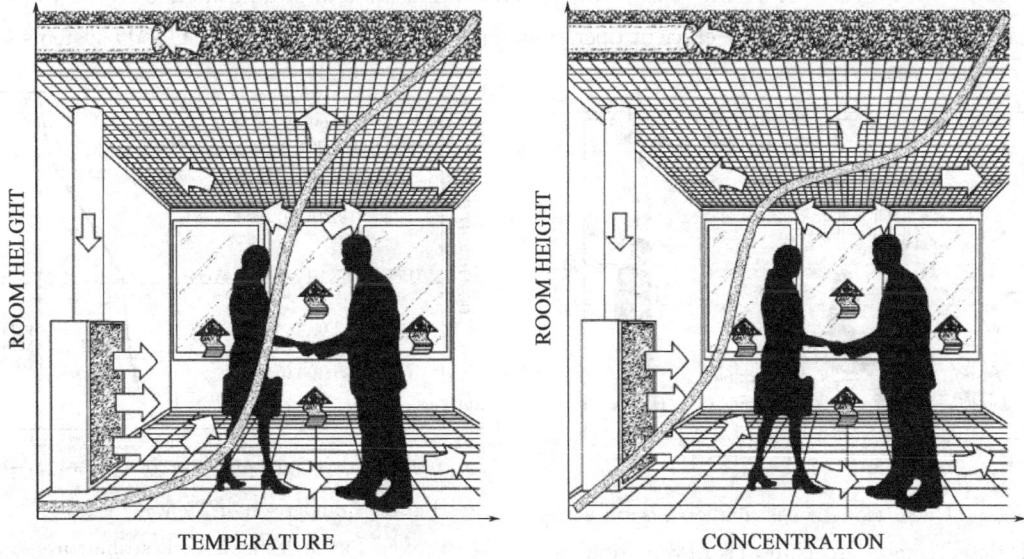

Fig. 3 Displacement Ventilation

Partially Mixed System

[19] Partially mixed room air distribution systems used for space cooling generally discharge conditioned air from a low sidewall or floor location, and the diffuser discharge turbulence is considerably greater than in fully stratified (TDV) systems. This creates a zone of high entrainment near the plane of discharge. A common example of partially mixed systems is under floor air distribution (UFAD) systems. UFAD systems differ from TDV systems primarily in the way air is delivered to the space: (1) air is supplied at higher

velocities through smaller supply outlets, and (2) local air supply conditions are generally under the control of occupants, allowing comfort conditions to be optimized. By introducing supply air with greater momentum, UFAD systems alter conditions in the lower region of the space by increasing the amount of mixing and reducing the temperature gradient. At higher elevations in the room, above the influence of supply outlets, overall airflow performance is similar to that of TDV systems. Based on recent experimental results and an extension of displacement theory, three distinct zones in the room can be used to describe the room air diffusion for UFAD systems.

[20] Figure 4 shows a schematic of typical airflow patterns in an UFAD system in an office environment. The diagram identifies two characteristic heights in the room that define the three zones in the room: (1) the throw height ($X50$) of the floor diffusers, and (2) the stratification height (SH), similar to that found in TDV systems. As shown, UFAD diffusers typically create adjacent zones that have excessive draft and cool temperatures, making long-term occupancy not recommended. When under direct individual control by the occupant, however, these local thermal conditions may be acceptable, and even desirable. Increased mixing in the occupied zone diminishes ventilation effectiveness, compared to TDV systems. In any case, control and optimization of stratification is crucial to system design and sizing, energy-efficient operation, and comfort performance of UFAD systems.

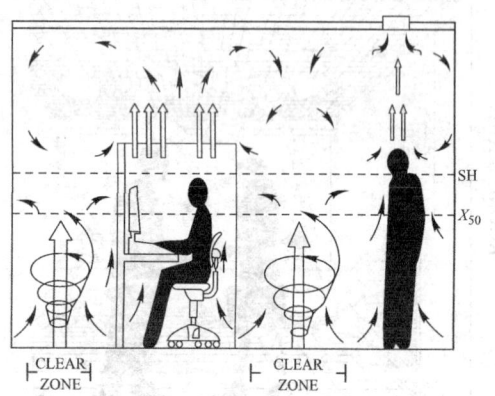

Fig. 4 Underfloor Air Distribution System with DiffuserTemperature Throw below Stratification Height

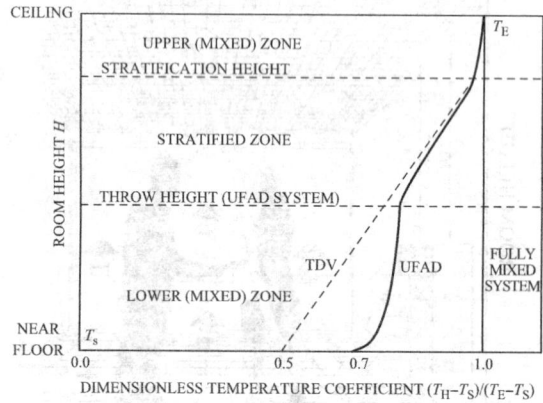

Fig. 5 Comparison of Typical Vertical Profiles for Underfloor Air Distribution, Displacement Ventilation and Mixing Systems

[21] Figure 5 compares typical vertical temperature profiles for UFAD, TDV, and conventional overhead mixing systems. The profiles shown are representative of normal operating conditions and are intended to demonstrate key differences and similarities between the three air distribution systems. The UFAD profile is based on temperatures in a space outside the direct influence of supply outlets (outside adjacent zones), and can vary significantly depending on several control factors (see the section on Controlling Stratification). In Figure 5, the no dimensional temperature (temperature ratio) is plotted versus

room height, where T_H is room air temperature as a function of height, T_S is supply temperature, and T_E is temperature at the ceiling. The linear profile for TDV systems is based on the 50% rule of thumb that applies to rooms of conventional height and normal heating loads; the temperature near the floor is assumed to be halfway between the supply and exhaust temperatures. The TDV profile is assumed to join the UFAD profile at the stratification height. **As long as the throw heights of the UFAD diffusers are below the stratification height, the upper zone is assumed to perform in a similar manner for both systems (for the same room-load-to-supply-volume ratio).** The fully mixed system profile represents a uniformly mixed room with the temperature equal to the exhaust temperature.

II. Words and Expressions

air-conditioning system	空气调节系统
humidity	湿度
cleanliness	洁净度
air distribution	气流分布
all-air system	全空气系统
air-water system	空气-水系统
all-water system	全水系统
direct expansion system	直接膨胀式系统
outdoor air intake	进风口
preheater	预热器
return air intake	回风口
filter	过滤器
dehumidifier	除湿装置
heating coil	加热盘管
humidifier	加湿器
duct system	管道系统
air outlet	排风口
air terminal	空气末端装置
refrigeration machine	制冷机组
pump	泵
cooling tower	冷却塔
boiler an auxiliaries	锅炉附件
environmental tobacco smoke	环境烟气量
ventilation effectiveness	通风效率
partially mixed system	部分混合系统
floor diffuser	地板送风口
thermal-displacement ventilation	换热通风
fully stratified systems	全分层系统

under-floor air distribution (UFAD) systems　　地板送风系统
supply and exhaust temperature　　送、排风温度

III. Notations

1. The comfort environment is the result of simultaneous control of temperature, humidity, cleanliness and air distribution with the occupant's vicinity that includes the proper acoustic level.

舒适性环境是居住者周围温度、湿度、空气洁净度以及气流分布同时调控的结果，并且居住者周围环境的噪声声级适度。

2. Heat is removed by a simple method using a finned tube coil with direct expansion of the refrigerant into the coil, hence, the term Dx. In general system, a Dx system can be described as a refrigeration system where the cooling effect is obtained directly from the refrigerant.

热量转移的一种简单方法就是利用一个翅片盘管让制冷剂在盘管中直接（吸热）膨胀，即直接膨胀。在一般的系统中，直膨式系统可以被当作制冷系统来描述，其冷却效应直接来源于制冷剂（蒸发吸热）。

3. Partially mixed systems provide some mixing within the occupied and/or process space while creating stratified conditions in the volume above. Most under-floor air distribution designs are examples of this type of system.

部分混合系统即散流器送风和空调房间或工艺过程的空调区域气流形成部分气流的混合，并产生气流分层。大部分地板送风气流组织设计都属于这类系统。

4. In practice, the actual temperature (or concentration) profile depends on the combined effects of various factors, but is largely driven by the characteristics of the room supply airflow and heat load configuration.

实际上，温度（或浓度）的分布情况是由多方面因素共同作用的，但在很大程度上还是由室内送风气流的特点和热负荷构成等所决定。

5. The U. S. Green Building Council's (USGBC 2005) Leadership in Energy and Environmental Design (LEED®) rating system, which was originally created in response to indoor air quality concerns, now includes prerequisites and credits for increasing ventilation effectiveness and improving thermal comfort.

美国联邦绿色建筑委员会制定的能源和环境设计（LEED）评价系统，最初只是针对室内空气品质，但是现已包含了提高室内通风效率、改善室内热舒适性的前提和保证。

6. For example, well-designed ceiling-based air distribution systems produce near-perfect air mixing in cooling mode, and yield an air change effectiveness of almost 1.0. Displacement and under-floor air distribution (UFAD) systems have the potential for values greater than 1.0.

例如，设计优良的顶棚送风系统在室内冷却过程中会产生近乎理想的气流混合效果，并且其气流置换效率能几乎达到1.0。置换通风系统和地板送风系统都有着很大的潜力，其通风效率都有可能大于1.0。

7. As long as the throw heights of the UFAD diffusers are below the stratification height, the upper zone is assumed to perform in a similar manner for both systems (for the same room-load-to-supply-volume ratio).

只要地板送风系统的散流器位置低于气流层高度,其上部空调空间的气流分布对两种系统都是相似的(空间负荷与送风量比相同情况下)。

IV. Exercises

1. Translate the following sentences into Chinese.

(1) All-air system is one in which the air is treated in a central refrigeration plant. The cold air is supplied to a space via ducts and distributed by means of terminal outlets mixing terminal outlets. There are no additional cooling loads at the treated space.

(2) Air-water system, in which the air apparatus and refrigeration plants are separate from the conditioned space. However, the cooling and heating of the conditioned space is affected in only a small part by air brought from the central apparatus. The major portion of the room thermal load is balanced by warm or cool water circulated either through a coil in an induction unit or through a radiant panel.

(3) All-water system are those with fan-coil room terminals to which are connected one or two water circuits. The cooling medium (such as chilled water or brine) may be supplied from a remote source and circulated through the coil of fan coil terminal, which is located in the conditioned space.

(4) Direct expansion system is widely used in a "packaged form", usually sized in small or intermediate tonnage ranges. Heat is removed by a simple method using a finned tube coil with direct expansion of the refrigerant into the coil, hence, the term Dx. In general system, a Dx system can be described as a refrigeration system where the cooling effect is obtained directly from the refrigerant.

(5) Environmental tobacco smoke (ETS) control is a LEED prerequisite. Banning indoor smoking is a common approach, but if indoor smoking is to be allowed, ANSI/ASHRAE *Standard* 62.1 requires that more than the base non-ETS ventilation air be provided where ETS is present in all or part of a building.

(6) The air change effectiveness is affected directly by the room air distribution system's design, construction, and operation, but is very difficult to predict. Many attempts have been made to quantify air change effectiveness, including ASHRAE *Standard* 129. However, this standard is only for experimental tests in well-controlled laboratories, and should not be applied directly to real buildings.

2. Translate the following sentences into English.

(1) 空气调节——实现对某一房间或空间的温度、湿度、洁净度和空气流速等进行调节和控制,并提供足够量的新鲜空气。空气调节简称空调,空调可以实现对建筑热湿环境、空气品质全面进行控制,或是说它包含了采暖和通风的部分功能。

(2) 全空气系统是完全由空气来担负房间冷热负荷的系统。一个全空气空调系统通过输送的冷空气向房间提供显热冷量和潜热冷量，其空气的冷却、去湿处理完全集中于空调机房内的空气处理机组来完成，在房间内不再进行补充冷却；而对输送到房间内的空气的加热可在空调机房内完成，也可以再个房间内完成。全空气系统的空气处理基本上集中于空调机房内完成，因此常称为集中空调系统。

（3）空气-水系统是由空气和水共同来承担室内冷、热负荷的系统，除了向室内送入经处理的新风，还在室内设有以水做介质的末端设备对室内空气进行冷却和加热。

Lesson 9　Boiler System（锅炉系统）

I. Text

[1] Boilers are used to supply steam or hot water for heating, processing, or power purposes. This chapter is primarily concerned with a description of the low-pressure steam and hot-water space heating boilers used in the heating systems of residences and small buildings.

[2] The basic construction of both low-pressure steam and hot-water space heating boilers fired by fossil fuels consists of an insulated steel jacket enclosing a lower chamber in which the combustion process takes place; and an upper chamber containing cast-iron sections or steel tubes in which water is heated or converted to steam for circulation through the pipes of the heating system.

Steam and Hot Water Boiler Similarities and Differences

[3] Steam and hot-water space heating boilers are very similar physically, but there are some important differences:

- Steam boilers operate only about three-fourths full of water, whereas hot-water boilers operate completely filled with water.
- Steam boilers in residential steam heating systems operate at 13.79 kPa pressure or slightly more, whereas residential hot-water boilers operate at approximately six times that pressure.
- Steam boilers are equipped with a low water cutoff device to protect the appliance from burning out if it should run out of water. Only large hot-water space heating boilers with a capacity exceeding 422.04 MJ/h are presently required by code to be equipped with low-water cutoffs. (Note: Many HVAC contractors who install the smaller residential hot-water boilers strongly recommend the addition of a low-water cutoff device to these appliances to prevent burn out if the boiler loses its water.)
- **Steam boilers require makeup feed to replace water lost through evaporation and the production of steam during normal operation.** Hot-water boilers can operate with little or no need for makeup water under the same normal operating conditions.

[4] The design and construction of the lower chamber depends upon the type of fuel used to fire the boiler. It serves as a combustion chamber for coal-fired and oil-fired boilers and as a compartment for housing the gas burner assembly on gas-fired boilers. **These gas burner assemblies are commonly designed for easy removal so that they can be periodically cleaned or serviced.**

[5] Oil burners are externally mounted with the burner nozzle extending into the combustion chamber. This is also true of gas conversion burners. Gas burner assemblies, on

the other hand, are located inside the lower chamber of the boiler.

[6] **The cast-iron sections or steel tubes in the upper chamber of the boiler contain water that circulates through the pipes in the heating system in the form of either steam or hot water.** The heat from the combustion process in the lower chamber of the boiler is transferred through the metal surface of the cast-iron sections or steel tubes to the water contained in them, causing a rise in temperature. **The amount of water contained in these passages is one of the ways in which steam boilers and hot-water space heating boilers are distinguished from one another.** In hot-water space heating boilers these passages are completely filled with water; whereas in low-pressure steam boilers only the lower two-thirds are filled. In steam boilers the water is heated very rapidly, causing steam to form in the upper one-third. The steam, under pressure, rises through the supply pipes connected to the top section of the boiler.

[7] A boiler jacket contains a number of different openings for pipe connections and the mounting of accessories. The number and type of openings on a specific boiler jacket depends upon the type of boiler (i. e. steam or hot water). Among the different openings to be found on a boiler jacket are the flue connections, water feed (supply) connection, inspection and cleanout tapping, blow down tapping, relief valve tapping, control tapping, drain tapping, expansion tank tapping, and return tapping. These are also gas and oil burner connections. Fig. 1 illustrates the arrangement of tapping in a Weil-McLain oil-fired boiler.

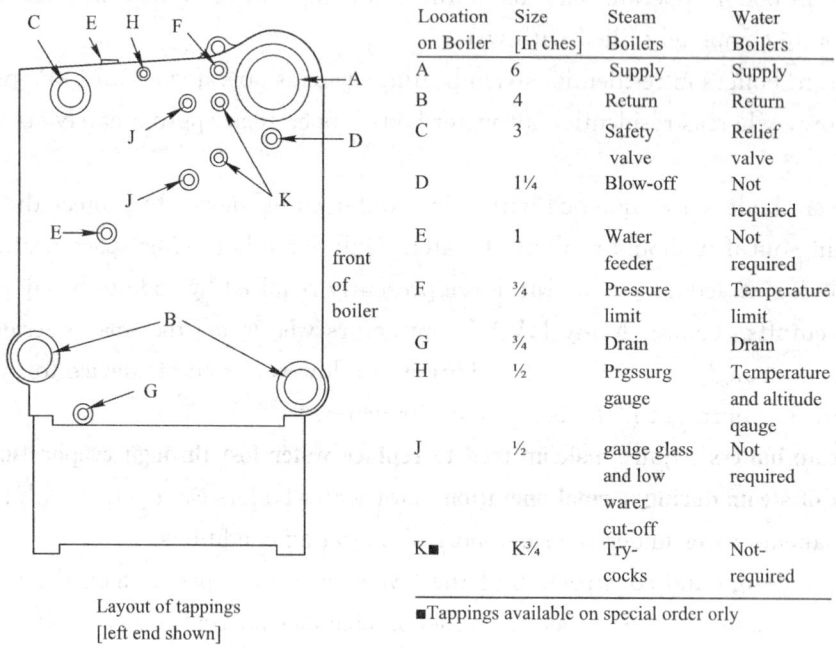

Location on Boiler	Size [Inches]	Steam Boilers	Water Boilers
A	6	Supply	Supply
B	4	Return	Return
C	3	Safety valve	Relief valve
D	1¼	Blow-off	Not required
E	1	Water feeder	Not required
F	¾	Pressure limit	Temperature limit
G	¾	Drain	Drain
H	½	Prgssurg gauge	Temperature and altitude qauge
J	½	gauge glass and low warer cut-off	Not required
K■	K¾	Try-cocks	Not-required

■Tappings available on special order orly

Layout of tappings [left end shown]

Fig. 1 Boiler control tapping locations

[8] **Most (but not all) of the controls on low-pressure steam and hot-water space heating boilers fired by the same fuel are similar in design and function, but there are exceptions.** For example, a few boiler controls and fittings are designed to be specifically used on steam

boilers; others are found only on hot-water space heating boilers.

Boiler Rating Method

[9] **The construction of low-pressure steel and cast-iron heating boilers is governed by the requirements of the ASME Boiler and Pressure Vessel Code.** This is a nationally recognized code used by boiler manufacturers, and any boiler used in a heating installation should clearly display the ASME stamp. State and local codes are usually patterned after the ASME Code.

[10] The location of the identification symbols used by the ASME is specified by the code and determined by the type of boiler. For example, on a water-tube boiler, it appears on a head of the steam-outlet drum near and above the manhole opening. On vertical fire-tube boilers, the stamp bearing the identification symbol should appear on the shell above the fire door and handhole opening. Other types of boilers (e.g. Scotch marine and super heaters) have their own specified location for the identification symbol stamp.

[11] **The ASME Boiler and Pressure Vessel Code applies only to boiler construction, specifically to maximum allowable working pressures, not to its heating capacity.** A number of different methods are used to rate the heating or operating capacity of a boiler. The boiler manufacturers have developed their own ratings, but these are generally used along with rating methods available from several professional and trade associations.

[12] The Steel Boilers Institute no longer exists, but its SBI rating is still found on many existing steel boilers. The I=B=R (or IBR) logo was created by the now defunct Institute of Boiler and Radiator Manufacturers to indicate the gross output (s) at 100 percent firing rate for most sectional cast-iron boilers. The I=B=R rating logo is now used by the Hydronic Institute Division of the Gas Appliance Manufacturers Association (GAMA).

[13] **The Mechanical Contractors Association of America has devised a method for rating boilers not covered by ether the SBI or I=B=R codes. Finally, gas-fired boilers are rated in accordance with methods developed by the American Gas Association.**

[14] Other rating logos appearing on boilers and in their installation and operation manuals are the Underwriters Laboratories, Inc. (UL) and the Underwriter's Laboratories of Canada logos.

[15] In terms of its heating capacity, the rating of a boiler can be expressed in square feet of equivalent direct radiation (EDR) or thousands of Btu/h. Sometimes a boiler horsepower rating is also given, but this has proven to be misleading.

[16] **For steam boilers, 0.09 m^2 of equivalent direct radiation (EDR) is equal to the emission of 253.21 kJ/h. For a water boiler, 0.09 m^2 of EDR is considered equal to the emission of 158.26 kJ/h.**

[17] A boiler horsepower (bhp) is the evaporation of 15.66 kg of water into dry steam from and at 100 ℃. For rating purposes, 1 bhp is considered as the heat equivalent of 13.02 m^2 of steam radiation per hour. In some cases bhp ratings are obtained by dividing steam SBI ratings by 140.

[18] A boiler is rated according to its operating or heating capacity, but this rating will vary in accordance with the type of load used as the basis for the rating. The three types of connected loads used to determine the rating of a boiler are:

(1) Net load

(2) Design load

(3) Gross load

[19] Net load refers to the actual connected load of the heat-emitting units in the steam or hot-water heating system. Design load includes the net-load rating plus an allowance for piping heat loss. Finally, gross load will equal the net load and the piping heat loss, plus an additional allowance for the pickup load.

Boiler Heating Surface

[20] The boiler heating surface (expressed in square feet) is that portion of the surface of the heat transfer apparatus in contact with the fluid being heated on one side and the gas or refractory being cooled on the other side. The direct or radiant heating surface is the surface against which the fire strikes. The surface that comes in contact with the hot gases is called the indirect or convection surface.

[21] The heating capacity of any boiler is influenced by the amount and arrangement of the heating surface and the temperature on either side. The arrangement of the heating surface refers to the ratio of the diameter of each passage to its length, as well as its contour (straight or curved), cross-sectional shape, number of passes, and other design variables.

Boiler Efficiency

[22] **The boiler efficiency is the ratio of the heat output to the caloric value of the fuel. Boiler efficiency is determined by various factors including the type of fuel used, the method of firing, and the control settings.** For example, oil- and gas-fired boilers have boiler efficiencies ranging from 70 to 80 percent. A hand-fired boiler in which anthracite coal is used will have a boiler efficiency of 60 to 75 percent.

Boiler energy Efficiency

[23] Two government programs have been created within the last 20 years to rate the energy efficiency of different heating appliances such as furnaces, boilers, water heaters, and heat pumps. **These two programs are (1) the annual fuel utilization capacity (AFUE) program and (2) the Energy Star Certification program.**

[24] Annual Fuel Utilization Capacity (AFUE). The energy efficiency of an oil-, gas-, or coal-fired boiler is measured by its annual fuel utilization capacity (AFUE). The AFUE ratings for boilers manufactured today are listed in the boiler manufacturer's literature. Look for the EnerGuide emblem for the efficiency rating of that particular model. The higher the rating, the more efficient the boiler. The government has established a minimum rating for boilers of 78 percent. Mid-efficiency boilers have AFUE ratings ranging from 78 to 82 percent. High-efficiency (condensing) boilers have AFUE ratings ranging from 88 to 97 percent. Conventional (noncondensing) steam and hot-water space heating

boilers have AFUE ratings of approximately 60 to 65 Percent.

[25] Energy Star Certification. Energy Star is an energy performance rating system created in 1992 by U.S. Environmental Protection Agency (EPA) to identify and certify certain energy-efficient appliances. The goal is to give special recognition to companies who manufacture products that help reduce greenhouse gas emissions. **This voluntary labeling program was expanded by 1995 to include furnaces, boilers, heat pumps, and other HVAC equipment.** Both the Energy Star label and an AFUE rating are used to identify an energy-efficient appliance.

Types of Boilers

[26] The boilers used in low-pressure steam and hot-water space heating systems can be classified in a number of different ways. Some of the criteria used in classifying them are:

(1) Construction material
(2) Construction design
(3) Boiler position
(4) Number of passes of the hot gases
(5) Length of travel of the hot gases
(6) Type of heating surface
(7) Type of fuel used

[27] Most boilers are constructed of either cast iron or steel. A few are constructed from nonferrous materials such as aluminum. **Cast-iron boilers generally display a greater resistance to the corrosive effects of water than steel ones do, but the degree of corrosion in steel boilers can be significantly reduced by chemically treating the water.**

[28] The heating core of many boilers is formed by joining together a series of cast-iron sections either horizontally (so-called pancake construction) or vertically. In the horizontal cast-iron section design, the heating surface of each cast-iron section is exposed at right angles to the rising flue gases. The water travels in a zigzag path from section to section in a manner similar to the flow of water in a steel tube boiler.

[29] Steel boilers may be classified with respect to the relative position of water and hot gases in the tubular heating surface. In fire-tube boilers, for example, the hot gases pass within the boiler tubes while the water being heated circulates around them. In water tube boilers, the reverse is true. Flexible steel tubes are used in some boilers for the circulation of the water around the heat rising from the fire.

[30] A hot-water (hydronic) copper-fin tube operates on a different principle from the cast-iron and steel boilers. It is designed to transfer heat almost instantly to the water. Water flows across the boiler heat exchanger, picks up heat, and then moves through the pipes to the heat convectors, radiators, or panels.

Note If the water stops flowing while the burner is still running, heat will build up until the water flashes into steam and damages the boiler. This condition is similar to dry firing in cast-iron and steel boilers. It can be avoided by installing a flow switch in the path of the water. The switch turns off the burner when the water stops flowing.

[31] Boilers can also be classified according to the number of passes made by the hot gases (e. g. one pass, two passes, and three passes). The length of travel of the hot gasses is another method used for classifying boilers. **The efficiency of a boiler heating surface depends, in part, upon the ratio of the cross-sectional area of the passage to its length.**

[32] Among the various fuels used to fire boilers are oil, gas (natural and propane), coal, and coke. Conversion kits for converting a boiler from one gas to another are available from some manufacturers. Changing from coal (or coke) to oil or gas can be accomplished by using conversion chambers and making certain other modifications.

[33] Electricity can also be used to fire boilers. **One advantage in using electric-fired boilers is that the draft provisions required by boilers using combustible fuels is not necessary.** Unlike the boilers fired by fossil fuels (oil, gas, coal, etc.), electric boilers do not have an AFUE efficiency rating. They operate at almost 100 percent efficiency.

[34] The classification criteria described above are selective and limited to the more common types in use. Considering the multiplicity of boiler types and designs available, it is extremely difficult to establish a classification system suitable for all of them.

II. Words and Expressions

steam	n.	蒸汽，蒸汽动力
fossil fuel		化石燃料
circulation	n.	流通，循环；传播
evaporation	n.	蒸发（作用）
combustion chamber		燃烧室
nozzle	n.	管嘴，喷嘴
accessory	n.	附件，配件
flue	n.	烟道
inspection	n.	检查，视察
relief valve		安全阀
ASME	abbr.	American Society of Mechanical Engineers 美国机械工程师协会
stamp	n	标志，印记
	vt.	标出；铭记
identification	n.	鉴定，认出；认同
manhole	n.	人孔，检修孔
rate	n.	比率，率
association	n.	联合，结合
gross output (s)		总产量
direct radiation		直接辐射

boiler horsepower		锅炉马力
		（锅炉蒸发量单位，等于 15.61kg/h）
refractory	n.	耐火材料
	adj.	耐熔的，难熔炼的
anthracite	n.	无烟煤
furnace	n.	熔炉，火炉
noncondensing	adj.	不凝固的
Environmental Protection Agency (EPA)		环境保护局
nonferrous	adj.	不含铁的，非铁的
aluminum	n.	铝
	adj.	铝的
corrosive	adj.	腐蚀（性）的
zigzag	n.	锯齿形的线条
	adj.	锯齿形的，Z 字形的
in a manner		在一定程度上
propane	n.	丙烷
coke	n.	焦煤，焦炭
modification	n.	缓和，减轻；改变

III. Notations

1. Steam boilers require makeup feed to replace water lost through evaporation and the production of steam during normal operation.

蒸汽锅炉需要补充正常运转过程中蒸发和产生蒸汽的水分流失。

2. These gas burner assemblies are commonly designed for easy removal so that they can be periodically cleaned or serviced.

这些气体燃烧器组件通常设计得便于拆装，使它们能定期清洁或维修。

3. The cast-iron sections or steel tubes in the upper chamber of the boiler contain water that circulates through the pipes in the heating system in the form of either steam or hot water.

铸铁部分或锅炉顶室的钢管含有水，这些水在加热系统中以蒸汽或热水的形式通过管道循环。

4. The amount of water contained in these passages is one of the ways in which steam boilers and hot-water space heating boilers are distinguished from one another.

这些通道中的含水量是蒸汽锅炉和热水采暖锅炉的一个不同点。

5. Most (but not all) of the controls on low-pressure steam and hot-water space heating boilers fired by the same fuel are similar in design and function, but there are exceptions.

大多数（但不是所有的）采用相同燃料的低压蒸汽、热水采暖锅炉的控制在设计和功能上是相似的，但也有例外。

6. The construction of low-pressure steel and cast-iron heating boilers is governed by

the requirements of the ASME Boiler and Pressure Vessel Code.

低压钢和铸铁采暖锅炉按 ASME 锅炉和压力容器规范的要求建造。

7. The ASME Boiler and Pressure Vessel Code applies only to boiler construction, specifically to maximum allowable working pressures, not to its heating capacity.

ASME 锅炉和压力容器规范仅适用于锅炉的建造，特别是最大允许工作压力，而不是它的供热能力。

8. The Mechanical Contractors Association of America has devised a method for rating boilers not covered by ether the SBI or I=B=R codes. Finally, gas-fired boilers are rated in accordance with methods developed by the American Gas Association.

美国机械师协会已经制定了一套既不覆盖 SBI 也不覆盖 IBR 规范的评价锅炉等级的方法。最终认为燃气锅炉的评价与美国天然气协会提出的方法相一致。

9. For steam boilers, 1 square foot of equivalent direct radiation (EDR) is equal to the emission of 240 Btu/h. For a water boiler, 1 square foot of EDR is considered equal to the emission of 150 Btu/h.

对于蒸汽锅炉，相当于直接辐射的 1 平方英尺为每小时 240Btu 的排放量，对于热水锅炉，相当于直接辐射的 1 平方英尺为每小时 150Btu 的排放量。

10. Boiler efficiency is determined by various factors including the type of fuel used, the method of firing, and the control settings.

锅炉效率由各种因素确定，这些因素包括使用燃料种类、燃烧方法以及控制设置。

11. These two programs are (1) the annual fuel utilization capacity (AFUE) program and (2) the Energy Star Certification program.

这两个程序分别是：(1) 年燃料利用率程序，(2) 能量最佳认证程序。

12. This voluntary labeling program was expanded by 1995 to include furnaces, boilers, heat pumps, and other HVAC equipment.

这种自愿标签计划扩展到 1995 年包括熔炉、锅炉、热泵和其他暖通空调设备。

13. Cast-iron boilers generally display a greater resistance to the corrosive effects of water than steel ones do, but the degree of corrosion in steel boilers can be significantly reduced by chemically treating the water.

铸铁锅炉比钢铁锅炉普遍表现出更大的抗水的腐蚀作用，但是钢铁锅炉的腐蚀程度可以通过化学处理水大大减少。

14. The efficiency of a boiler heating surface depends, in part, upon the ratio of the cross-sectional area of the passage to its length.

锅炉受热面的效率在某种程度上取决于通道的截面积与长度的比例。

15. One advantage in using electric-fired boilers is that the draft provisions required by boilers using combustible fuels are not necessary.

电动燃油锅炉的一个优点是，使用燃烧燃料的锅炉必须的规范对于电动燃油锅炉则是不必要的。

IV. Exercises

1. Translate the following sentences into Chinese.

(1) The basic construction of both low-pressure steam and hot-water space heating boilers fired by fossil fuels consists of an insulated steel jacket enclosing a lower chamber in which the combustion process takes place; and an upper chamber containing cast-iron sections or steel tubes in which water is heated or converted to steam for circulation through the pipes of the heating system.

(2) Steam boilers are equipped with a low-water cutoff device to protect the appliance from burning out if it should run out of water. Only large hot-water space heating boilers with a capacity exceeding 400,000 Btu/h are presently required by code to be equipped with low-water cutoffs.

(3) In hot-water space heating boilers these passages are completely filled with water; whereas in low-pressure steam boilers only the lower two-thirds are filled. In steam boilers the water is heated very rapidly, causing steam to form in the upper one-third. The steam, under pressure, rises through the supply pipes connected to the top section of the boiler.

(4) Among the different openings to be found on a boiler jacket are the flue connections, water feed (supply) connection, inspection and cleanout tapping, blow down tapping, relief valve tapping, control tapping, drain tapping, expansion tank tapping, and return tapping.

(5) The Steel Boilers Institute no longer exists, but its SBI rating is still found on many existing steel boilers. The I=B=R (or IBR) logo was created by the now defunct Institute of Boiler and Radiator Manufacturers to indicate the gross output (s) at 100 percent firing rate for most sectional cast-iron boilers. The I=B=R rating logo is now used by the Hydronic Institute Division of the Gas Appliance Manufacturers Association (GAMA).

(6) A boiler horsepower (bhp) is the evaporation of 34.5 Ib of water into dry steam from and at 212°F. For rating purposes, 1 bhp is considered as the heat equivalent of 140 ft^2 of steam radiation per hour. In some cases bhp ratings are obtained by dividing steam SBI ratings by 140.

(7) The boiler heating surface (expressed in square feet) is that portion of the surface of the heat transfer apparatus in contact with the fluid being heated on one side and the gas or refractory being cooled on the other side. The direct or radiant heating surface is the surface against which the fire strikes. The surface that comes in contact with the hot gases is called the indirect or convection surface.

(8) The energy efficiency of an oil-, gas-, or coal-fired boiler is measured by its annual fuel utilization capacity (AFUE). The AFUE ratings for boilers manufactured today are listed in the boiler manufacturer's literature. Look for the EnerGuide emblem for the efficiency rating of that particular model.

(9) The heating core of many boilers is formed by joining together a series of cast-iron sections either horizontally (so-called pancake construction) or vertically. In the horizontal cast-iron section design, the heating surface of each cast-iron section is exposed at right angles to the rising flue gases. The water travels in a zigzag path from section to section in a manner similar to the flow of water in a steel tube boiler.

(10) If the water stops flowing while the burner is still running, heat will build up until the water flashes into steam and damages the boiler. This condition is similar to dry firing in cast-iron and steel boilers. It can be avoided by installing a flow switch in the path of the water. The switch turns off the burner when the water stops flowing.

2. Translate the following sentences into English.

(1) 锅炉是一种将化学能以及工业生产中的余热或其他热源，转化为一定温度和压力的水或蒸汽的换热设备。

(2) 按锅炉传热方式的特点，工业锅炉受热面可以分为辐射受热面和对流受热面。辐射受热面是布置在锅炉炉膛内吸收辐射热的那一部分受热面，主要是水冷壁受热面；对流受热面是布置在锅炉烟道中，受高温烟气直接冲刷并以对流方式传递热量的那一部分受热面。

(3) 锅炉热效率是锅炉的重要技术指标。锅炉热效率是指锅炉额定符合运行时，每小时送进锅炉的燃料全部完全燃烧所放出的热量中有百分之几用来产生蒸汽或加热水。

(4) 锅炉用途广泛，种类很多，常用的分类方法有：按锅炉结构分类、按用途分类、按所用燃料或能源分类、按燃烧方式分类、按通风方式分类、按循环方式分类以及按锅筒布置形式分类等。

Lesson 10 Refrigeration Systems（制冷系统）

I. Text

[1] The most widely used method of producing mechanical refrigeration is called the vapor compression refrigeration system. In this chapter we will explain how refrigeration is accomplished by this method and what equipment is required for it. Some of the basic calculations used in determining the performance of the system will also be introduced.

[2] **The equipment arrangement and interconnecting piping for the basic vapor compression system is shown in Figure 1.** Typical operating conditions have been selected in order to make the discussion more practical.

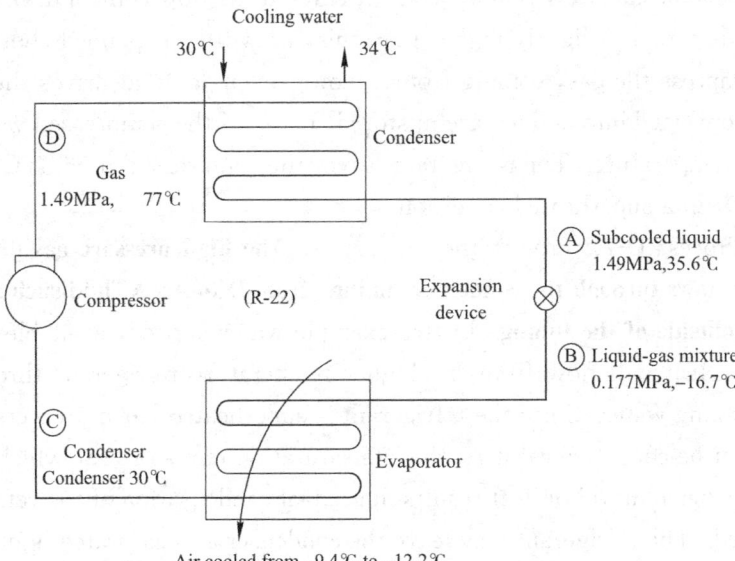

Fig. 1 Basic vapor compression refrigeration system with example of possible operating conditions

[3] The four basic components of the system are the expansion device (also called the flow control device), evaporator, compressor, and condenser.

[4] The Process A-B through the Flow Control Device. Liquid refrigerant R22 at 1.47 MPa (gage pressure) and 35.6℃ enters the expansion device, point A. There are various types of expansion devices; an expansion valve or capillary tube is two common types. In all cases the flow control has a narrow opening, which results in a large pressure loss as the refrigeration flows through it. The refrigerant leaves at point B at 0.18 MPa. Because this pressure is below the saturation pressure corresponding to 35.6 ℃, some of the liquid refrigerant immediately flashes to gas. The portion of the liquid that evaporates takes the latent heat required for its evaporation from the flowing mixture, thus cooling it.

The refrigerant leaves the valve as a liquid-vapor mixture in the saturated state. The saturation temperature for R22 at 0.18 MPa is −16.7 ℃; therefore this will be the refrigerant temperature at B.

[5] The Process B-C through the Evaporator. The refrigerant flows through the evaporator tubing from B to C. The substance to be cooled, usually air or a liquid, flows over the outside of the tubes. It is at a temperature higher than the refrigerant in the evaporator, therefore heat will flow from it through the tube wall to the refrigerant. In the case we have chosen, air is cooled from −9.4 ℃ to −12.2 ℃. **Because the liquid refrigerant in the evaporator is already at its saturation temperature the heat that it gains will cause it to evaporate as it travels through the evaporator.** The refrigerant generally leaves the evaporator either as a saturated vapor or superheated vapor.

[6] The Process C-D through the Compressor. The compressor draws the vapor into its suction side and then compresses it to a suitable high pressure for condensing. This pressure will be approximately that at which it entered the flow control device, 1.48 MPa. (The pressure is actually slightly higher than this, as will be explained shortly). Work is required to compress the gas, coming from a motor or engine that drives the compressor. This work is converted into an increase in stored energy of the compressed vapor, resulting in a rise in its temperature. The refrigerant leaves the compressor at 76.7 ℃ in this example, at point D, in a superheated condition.

[7] The Process D-A through the Condenser. **The high pressure gas discharged from the compressor flows through the condenser tubing, from D to A. A fluid such as air or water flows over the outside of the tubing.** In this example water is used, available at a temperature of 30 ℃. Heat will flow from the higher temperature refrigerant through the tube walls to the cooling water. Since the refrigerant is superheated when it enters the condenser it will at first be cooled, until it reaches its saturation temperature, which at 1.47 MPa is 41.1 ℃. Further removal of heat results in gradual condensation of the refrigerant, until it is all liquefied. The refrigerant may leave the condenser as a saturated liquid or it may be sub-cooled. In this example we assumed it was sub-cooled to 35.6 ℃ before entering the flow control device.

[8] It is essential to be able to determine the refrigeration system's performance. Some of the performance characteristics that are of concern are the cooling (refrigeration) capacity, the power required for the compressor, the refrigerant flow rate, and the rate of heat rejected (removed) in the condenser.

The Refrigeration Effect

[9] **It is called the refrigeration effect because it is also the amount of heat removed from the medium to be cooled, for each pound or kilogram of refrigerant flowing.** This follows from the Energy Equation. That is

$$R.E. = h_c - h_b = h_c - h_a \tag{1}$$

where

$R.E$ = refrigeration effect, kJ/kg

h_c = enthalpy of refrigerant leaving evaporator, kJ/kg

h_b, h_a = enthalpy of refrigerant entering evaporator, kJ/kg

[10] Note that the value of the enthalpy entering the evaporator, h_b, is the same as the value entering the flow control device, h_a. This is true because the process A-B is at constant enthalpy. For this reason both h_c and h_a can be read from the saturation tables rather than from the p-h diagram, for better accuracy, as seen in the following example.

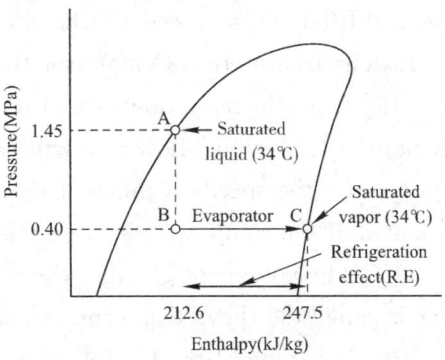

Fig. 2 Ideal cycle evaporator process, B-C, constant pressure.

Refrigerant Mass Flow Rate

[11] The refrigerant mass flow rate circulating through a system to produce given refrigeration capacity can be found as follows:

$$m = \frac{Q}{R.E.} \qquad (2)$$

where

m = mass flow rate, kg/s

Q = system refrigeration capacity, kJ/s

Required Theoretical Compressor Power

[12] **It is usually more valuable to determine how much power is needed to drive the compressor, rather than the work required.** This can be found from the work of compression and the mass flow rate, using the following equation:

$$P = W \times m \qquad (3)$$

where

P = compressor required theoretical power, kJ/s

W = work (heat) of compression, kJ/kg

m = mass flow rate, kg/s

[13] The compressor power is more conveniently expressed in units of horsepower rather than kJ/s. It is also useful to determine the required power in horsepower per ton of refrigeration, in order to compare the effect of operating at different conditions.

[14] The power required to drive the compressor in the ideal cycle is called the theoretical power. A very important fact concerning this power is:

[15] **The minimum theoretical power required to drive the compressor occurs in the ideal cycle, for any given conditions.**

[16] The importance of this is that the power required for an actual system can be measured and compared to the best possible situation—the ideal cycle. This provides a goal for purposes of minimizing energy consumption. This goal can never be reached, but it

provides a good frame of reference.

[17] The minimum power is required because the compression is an isentropic (adiabatic and frictionless) process. The proof of this can be found in thermodynamics textbook.

Required Theoretical Compressor Displacement

[18] After the mass flow rate of refrigerant is determined, the volume flow rate can be calculated. The volume flow rate will vary, depending on where in the system it is determined, since the specific volume of the refrigerant varies. Usually the volume flow rate is calculated at the compressor suction inlet.

[19] The volume of gas that the compressor must be capable of handling in the ideal cycle is called the theoretical compressor displacement.

[20] It is found from the following equation:
$$V_t = v \times m \tag{4}$$
where

V_t = theoretical compressor displacement, m^3/s

v = specific volume of refrigerant at compressor suction, m^3/kg

m = mass flow rate of refrigerant, kg/s

[21] The required compressor displacement for the ideal cycle is called the theoretical displacement because it is the minimum possible displacement.

The Coefficient of Performance

[22] It is useful to have a single measurement that describes how effectively a refrigeration machine is performing. The coefficient of performance (COP) serves this purpose. It is defined as:
$$COP = \frac{refrigeration\ capacity\ (Q_e)}{net\ power\ input\ (P),\ in\ same\ units} \tag{5}$$

[23] In this equation the refrigeration capacity of the system, Q_e, and the net power input to the compressor, P must be expressed in the same units For example, if the refrigeration capacity is expressed in Btu per hour then the power input must also be expressed in Btu per hour.

[24] The COP provides a measurement of the energy use efficiency of the system. Because we always wish to have the largest refrigeration capacity with the smallest expenditure of power, the largest practical valve of the COP is desirable. The COP can also be expressed in terms of the units used in the thermodynamic cycle of the vapor compression system. In this case it is :
$$COP = \frac{refrigeration\ effect}{heat\ of\ compression} \tag{6}$$

[25] The units of both terms must still be the same, such as Btu/lb or kJ/kg.

[26] It should be understood that Equation (5) is also the definition of the COP for any refrigeration system, regardless of how the refrigeration is accomplished. The definition is valid for an absorption system as well as a vapor compression system. Equation (6)

is simply the expression of the *COP* for the case of the vapor compression cycle.

[27] The *COP* of real systems is always less than that of ideal cycles, due to unavoidable losses such as friction.

[28] It is possible to determine the maximum possible *COP* for any given evaporating and condensing temperatures. This value is even greater than that for the ideal vapor compression cycle. This is discussed in detail later in this chapter. Another performance factor similar to the *COP*, called the energy efficiency ratio (*EER*), which is now widely used, is also discussed later.

Energy Conservation

[29] A summary of cycle effects that result in reduced energy consumption per unit of refrigeration capacity are listed here. In some cases suggestions of how to achieve these benefits will be postponed to a more appropriate place.

1. Operate at low condensing temperatures. This can be achieved by using large condensers and maintaining clean heat transfer surfaces. (It will be explained later that an extremely low condensing temperature may cause erratic system performance.)

2. Operate at high evaporating temperatures. This can be achieved by using large evaporators.

3. Size refrigerant pipe lines for reasonably low pressure drops.

4. Design and operate the system to provide significant liquid sub-cooling in the condenser.

5. Use a liquid-suction heat exchanger if analysis shows it improves cycle performance significantly, but do not use with a refrigerant where the hot gas discharge temperature would be too high.

The Energy Efficiency Ratio (*EER*)

[30] Another measure of efficiency of performance of refrigeration equipment, other than the *COP*, is the energy efficiency ratio (*EER*). It is expressed by the following equation:

$$EER = \frac{Q_e}{P} = \frac{useful\ cooling\ capacity}{power\ input} \tag{7}$$

[31] The *EER* has the same two terms as the equation for the *COP* has, and therefore it measures the same efficiency use of energy. However, the units in the equation differ, and therefore the numerical values of the *EER* will differ from these of the *COP* for the same conditions. The *EER* has been developed because it is easier for the consumer to understand and use. Labeling of air conditioning and refrigeration equipment with *EER* values at a standard set of conditions is legally required in certain places and circumstances. Since the *EER* (and *COP*) changes with conditions, it is often difficult to obtain a realistic *EER* under varying operating conditions. One attempt to accomplish this is by use of the Seasonal Energy Efficiency Ratio (*SEER*), which tries to measure the average *EER* of equipment for a cooling season.

Maximum Coefficient of Performance

[32] It can be shown from the Second Law of Thermodynamics that a refrigeration

system has a maximum possible coefficient of performance. The equation expressing this is:

$$COP_m = \frac{T_1}{T_2 - T_1} \tag{8}$$

where

COP_m = maximum possible COP for a refrigeration system
T_1 = temperature at which heat is absorbed from cooling load
T_2 = temperature at which heat is rejected to heat sink

[33] The derivation of this equation can be found in thermodynamics textbook. Temperatures in the equation must be expressed in absolute units, Kelvin (K), or Rankine (R).

[34] The value of Equation (8) is that it shows the upper limit of efficiency. In reality, the COP of real machines is always considerably less than the maximum possible, due to friction and other losses.

II. Words and Expressions

device	*n.*	装置
capillary tube	*n.*	毛细管
liquid-vapor mixture	*n.*	气液混合物
saturation temperature	*n.*	饱和温度
saturate	*v.*	使渗透，浸，使饱和
	adj.	浸透的，饱和度高的
suction side		吸入侧
discharge	*v.*	卸下，放出
characteristic	*adj.*	特有的，典型的
	n.	特性，特征，特色
cooling (refrigeration) capacity		冷却能力
reject	*v.*	除去，滤去
accuracy	*n.*	准确（性），精确度
circulate	*v.*	流通，循环，传播
refrigerant mass flow rate		制冷剂质量流率
the ideal cycle		理想循环
isentropic	*n.*	等熵线
	adj.	等熵的
volume flow rate		体积流率
theoretical compressor displacement		理论压缩机位移
the coefficient of performance		性能系数
the thermodynamic cycle		热力循环
an absorption system		吸收式系统
friction	*n.*	摩擦，摩擦力
in detail		详细的

erratic	*adj.*	不稳定的，偏离的
the energy efficiency ratio		能效比
equipment	*n.*	设备

III. Notations

1. The equipment arrangement and interconnecting piping for basic vapor compression system is shown in Figure 1.

图 1 示出了蒸汽压缩系统的设备布置及管道连接。

2. Because the liquid refrigerant in the evaporator is already at its saturation temperature the heat that it gains will cause it to evaporate as it travels through the evaporator.

由于蒸发器中的液态制冷剂已经处于饱和温度下，在流过蒸发器时获得的热量使其蒸发。

3. The high pressure gas discharged from the compressor flows through the condenser tubing, from D to A. A fluid such as air or water flows over the outside of the tubing.

从压缩机中放出的高压气体进入冷凝器管道流动，即从状态 D 到 A。一种流体如空气或水在管外侧流动。

4. It is called the refrigeration effect because it is also the amount of heat removed from the medium to be cooled, for each pound or kilogram of refrigerant flowing.

每英镑或千克的制冷剂流从被冷却介质吸取的热量称作制冷量。

5. It is usually more valuable to determine how much power is needed to drive the compressor, rather than the work required.

通常确定压缩机运行所需的功率而不是功更有意义。

6. The minimum theoretical power required to drive the compressor occurs in the ideal cycle, for any given conditions.

对任何给定工况，压缩机运行所需的最小理论功率出现在理想循环中。

IV. Exercises

1. Translate the following sentences into Chinese.

(1) In all cases the flow control has a narrow opening, which results in a large pressure loss as the refrigeration flows through it. The refrigerant leaves at point B at 0.18 MPa. Because this pressure is below the saturation pressure corresponding to 35.6℃, some of the liquid refrigerant immediately flashes to gas.

(2) Work is required to compress the gas, coming from a motor or engine that drives the compressor. This work is converted into an increase in stored energy of the compressed vapor, resulting in a rise in its temperature. The refrigerant leaves the compressor at 76.7℃ in this example, at point D, in a superheated condition.

(3) Further removal of heat results in gradual condensation of the refrigerant, until it is all liquefied. The refrigerant may leave the condenser as a saturated liquid or it may be sub-cooled. In this example we assumed it was sub-cooled to 35.6℃ before entering the flow control device.

(4) The compressor power is more conveniently expressed in units of horsepower rather than Btu per minute. It is also useful to determine the required power in horsepower per ton of refrigeration, in order to compare the effect of operating at different conditions.

(5) The importance of this is that the power required for an actual system can be measured and compared to the best possible situation—the ideal cycle. This provides a goal for purposes of minimizing energy consumption. This goal can never be reached, but it provides a good frame of reference.

(6) In this equation the refrigeration capacity of the system, Q_e, and the net power input to the compressor, P must be expressed in the same units. For example, if the refrigeration capacity is expressed in Btu per hour then the power input must also be expressed in Btu per hour.

(7) It should be understood that Equation (5) is also the definition of the COP for any refrigeration system, regardless of how the refrigeration is accomplished. The definition is valid for an absorption system as well as a vapor compression system. Equation (6) is simply the expression of the COP for the case of the vapor compression cycle.

(8) A summary of cycle effects that result in reduced energy consumption per unit of refrigeration capacity are listed here. In some cases suggestions of how to achieve these benefits will be postponed to a more appropriate place.

(9) The EER has the same two terms as the equation for the COP has, and therefore it measures the same efficiency use of energy. However, the units in the equation differ, and therefore the numerical values of the EER will differ from these of the COP for the same conditions. The EER has been developed because it is easier for the consumer to understand and use.

2. Translate the following sentences into English.

(1) 蒸气压缩式制冷系统由压缩机、冷凝器、膨胀阀、蒸发器组成，用管道将其连成一个封闭的系统。工质在蒸发器内与被冷却对象发生热量交换，吸收被冷却对象的热量并汽化，产生的低压蒸气被压缩机吸入，经压缩后以高压排出。

(2) 在整个循环过程中，压缩机起着压缩和输送制冷剂蒸气并造成蒸发器中低压力、冷凝器中高压力的作用，是整个系统的心脏；节流阀对制冷剂起节流降压作用并调节进入蒸发器的制冷剂流量；蒸发器是输出冷量的设备，制冷剂在蒸发器中吸收被冷却物体的热量，从而达到制取冷量的目的；冷凝器是输出热量的设备，从蒸发器中吸取的热量连同压缩机消耗的功所转化的热量在冷凝器中被冷却介质带走。

(3) 液体蒸发式制冷中，制冷剂在要求的低温下蒸发，从被冷却对象中吸取热量；再在较高的温度下凝结，向外界排放热量。所以，只有在工作温度范围能够汽化和凝结的物质才有可能作为制冷剂使用。多数制冷剂在常温和常压下呈气态。

(4) 混合制冷剂是由两种或两种以上纯制冷剂组成的混合物。由于纯制冷剂在品种和性质上的局限性，采用混合物做制冷剂为调制制冷剂的性质和扩大制冷剂的选择方面提供了更大的自由度。

Lesson 11   Description of Ground-Source Types for Heat Pump (地源热泵)

I. Text

[1] The ground system links the heat pump to the underground and allows for extraction of heat from the ground or injection of heat into the ground. These systems can be classified generally as open or closed systems, with a third category for those not truly belonging to one or the other.

• Open systems: Groundwater is used as a heat carrier, and is brought directly to the heat pump. Between rock/soil, ground water, and the heat pump evaporator is no barrier; hence this type is called "open".

• **Closed systems: Heat exchangers are located in the underground (either in a horizontal, vertical or oblique fashion), and a heat carrier medium is circulated within the heat exchangers, transporting heat from the ground to the heat pump (or vice versa)**. The heat carrier is separated from the rock/soil and groundwater by the wall of the heat exchanger, making it a "closed" system.

• Other systems: Not always the system can be attributed exactly to one of the above categories, e. g. , if there is a certain distinction between groundwater and the heat carrier fluid, but no true barrier. Standing column wells, mine water or tunnel water are examples for this category.

[2] **To choose the right system for a specific installation, several factors have to be considered**: geology and hydrogeology of the underground (sufficient permeability is a must for open systems), area and utilisation on the surface (horizontal closed systems require a certain area), existence of potential heat sources like mines, and the heating and cooling characteristics of the building (s) . In the design phase, more accurate data for the key parameters for the chosen technology are necessary; to size the ground system in such a way that optimum performance is achieved with minimum cost. The individual types of ground systems are described in more detail as following.

Closed Systems

Horizontal-loop Systems

[3] The closed system easiest to install is the horizontal ground heat exchanger (synonym: ground heat collector, horizontal loop). Due to restrictions in the area available, in Western and Central Europe the individual pipes are laid in a relatively dense pattern, connected either in series or in parallel (Fig. 1).

[4] For the ground heat collectors with dense pipe pattern, usually the top earth layer

is removed completely, the pipes are laid, and the soil is distributed back over the pipes. In Northern Europe (and in North America), where land area is cheaper, a wide pattern ("loop") with pipes laid in trenches is preferred (Fig. 2). Trenching machines facilitate installation of pipes and backfilling.

[5] To save surface area with ground heat collectors, some special ground heat exchangers have been developed. Exploiting a smaller area at the same volume, these collectors are best suited for heat pump systems for heating and cooling, where natural temperature recharge of the ground is not vital. Hence these collectors are widely used in Northern America, and one type only, the trench collector (Fig. 3), achieved a certain distribution in Europe, mainly in Austria and Southern Germany. For the trench collector, a number of pipes with small diameter are attached to the steeply inclined walls of a trench some meters deep.

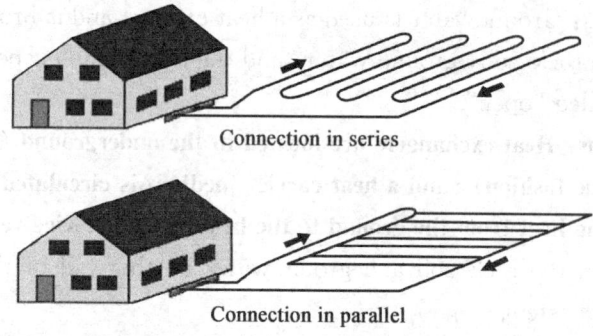

Fig. 1　Horizontal ground heat exchanger (European style)

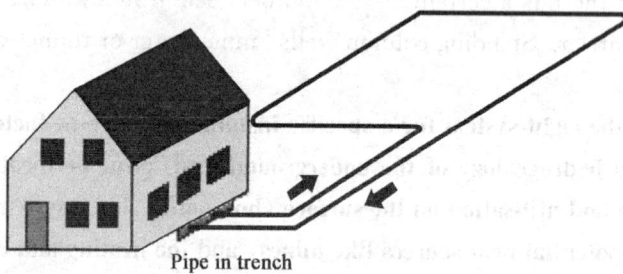

Fig. 2　Horizontal ground heat exchanger (North European and American style)

[6] The main thermal recharge for all horizontal systems is provided for mainly by the solar radiation to the earth's surface. It is important not to cover the surface above the ground heat collector, or to operate it as a heat store, if it has to be located e. g. , under a building. A variation of the horizontal GSHP is direct expansion. **In this case, the working medium of the heat pump (refrigerant) is circulating directly through the ground heat collector pipes (in other words, the heat pump evaporator is extended into the ground)**. The advantage of this technology is the omission of one heat exchange process, and thus a possibility for better system efficiency. In France and Austria, direct expansion also has been coupled to direct condensation in the floor heating system. DX (direct-expansion) requires good knowledge of the refrigeration cycle, and is restricted to smaller units. The horizontal-

loop systems can be buried beneath lawns, landscaping, and parking lots. Horizontal systems tend to be more popular where there is ample land area with a high water table.

• Advantages: Trenching costs typically lower than well-drilling costs; flexible installation options.

• Disadvantages: Large ground area required; ground temperature subject to seasonal variance at shallow depths; thermal properties of soil fluctuate with season, rainfall, and burial depth; soil dryness must be properly accounted for in designing the required pipe length, especially in sandy soils and on hilltops that may dry out during the summer; pipe system could be damaged during backfill process; longer pipe lengths are required than for vertical wells; antifreeze solution viscosity increases pumping energy, decreases the heat-transfer rate, and thus reduces overall efficiency; lower system efficiencies.

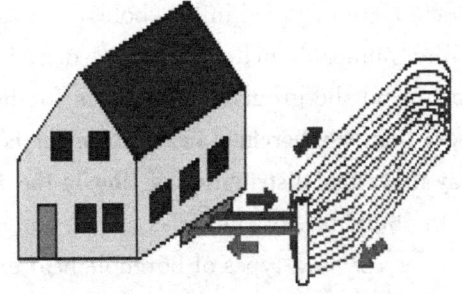

Fig. 3   Trench collector

Spiral Loops

[7] A variation on the multiple pipe horizontal-loop configuration is the spiral loop, commonly referred to as the "slinky". The spiral loop consists of pipe unrolled in circular loops in trenches and the horizontal configuration. Another variation of the spiral-loop system involves placing the loops upright in narrow vertical trenches. The spiral-loop configuration generally requires more piping, typically 500 to 1000 ft per system cooling ton (43.3 to 86.6 m/kW) but less total trenching than the multiple horizontal-loop systems described above. For the horizontal spiral-loop layout, trenches are generally 3 to 6 ft (0.9 to 1.8 m) wide; multiple trenches are typically spaced about 12 ft (3.7 m) apart. For the vertical spiral-loop layout, trenches are generally 6 in (15.2 cm) wide; the pipe loops stand vertically in the narrow trenches. In cases where trenching is a large component of the overall installation costs, spiral-loop systems are a means of reducing the installation cost. **As noted with horizontal systems, slinky systems are also generally associated with lower tonnage systems where land area requirements are not a limiting factor.**

• Advantages: Requires less ground area and less trenching than other horizontal-loop designs; installation costs sometimes less than other horizontal-loop designs.

• Disadvantages: Requires more total pipe length than other ground coupled designs; relatively large ground area required; ground temperature subject to seasonal variance; larger pumping energy requirements than other horizontal loops defined above; backfilling the trench can be difficult with certain soil types and the pipe system could be damaged during backfill process.

Vertical Loops

[8] As can be seen from measurements dating as far back as to the 17th century, the

temperature below a certain depth ("neutral zone", at ca. 15~20 m depth) remains constant over the year. This fact, and the need to install sufficient heat exchange capacity under a confined surface area, favours vertical ground heat exchangers (borehole heat exchangers). In a standard borehole heat exchanger, plastic pipes (polyethylene or polypropylene) are installed in boreholes, and the remaining room in the hole is filled (grouted) with a pumpable material. In Sweden, boreholes in hard, crystalline rock usually are kept open, and the groundwater serves for heat exchange between the pipes and the rock. **If more than one borehole heat exchanger is required, the pipes should be connected in such a way that equal distribution of flow in the different channels is secured.** Manifolds can be in or at the building, or the pipes can be connected in trenches in the field (Fig. 4).

[9] Several types of borehole heat exchangers have been used or tested; the two possible basic concepts are:

• U-pipes, consisting of a pair of straight pipes, connected by a 180l turn at the bottom. One, two or even three of such U-pipes are installed in one hole. The advantage of the U-pipe is low cost of the pipe material, resulting in double U pipes being the most frequently used borehole.

• Coaxial (concentric) pipes, either in a very simple way with two straight pipes of different diameter, or in complex configurations heat exchangers in Europe.

[10] Vertical loops are generally considered when land surface is limited. Wells are bored to depths that typically range from 75 to 300 ft (22.9 to 91.4 m) deep. The closed-loop pipes are inserted into the vertical well. Typical piping requirements range from 200 to 600 ft per system cooling ton (17.4 to 52.2 m/kW), depending on soil and temperature conditions.

[11] Multiple wells are typically required with well spacing not less than 15 ft (4.6 m) in the northern climates and not less than 20 ft (6.1 m) in southern climates to achieve the total heat-transfer requirements. A 300~500 ton capacity system can be installed on 1 acre of land, depending on soil conditions and ground temperature. There are three basic types of vertical-system heat exchangers: U-tube, divided-tube, and concentric-tube (pipe-in-pipe) system configurations.

• Advantages: Requires less total pipe length than most closed-loop designs; requires the least pumping energy of closed-loop systems; requires least amount of surface ground area; ground temperature typically not subject to seasonal variation.

• Disadvantages: Requires drilling equipment; drilling costs frequently higher than horizontal trenching costs; some potential for long-term heat build-up underground with inadequately spaced boreholes.

Submerged Loops

[12] If a moderately sized pond or lake is available, the closed-loop piping system can be submerged. Some companies have installed ponds on facility grounds to act as ground-coupled systems; ponds also serve to improve facility aesthetics. **Submerged-loop applica-**

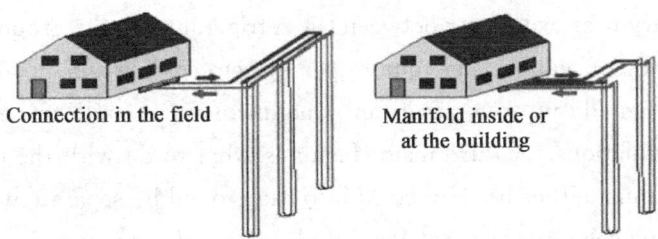

Fig. 4 Borehole heat exchangers (double-U-pipe)

tions require some special considerations, and it is best to discuss these directly with an engineer experienced in the design applications. This type of system requires adequate surface area and depth to function adequately in response to heating or cooling requirements under local weather conditions.

[13] In general, the submerged piping system is installed in loops attached to concrete anchors. Typical installations require around 300 ft of heat-transfer piping per system cooling ton (26.0 m/kW) and around 3000 ft^2 of pond surface area per ton (79.2 m^2/kW) with a recommended minimum one-half acre total surface area. The concrete anchors act to secure the piping, restricting movement, but also hold the piping 9 to 18 in (22.9 to 45.7 cm) above the pond floor, allowing for good convective flow of water around the heat-transfer surface area.

[14] It is also recommended that the heat-transfer loop be at least 6 to 8 ft (1.8 to 2.4 m) below the pond surface, preferably deeper. This maintains adequate thermal mass even in times of extended drought or other low-water conditions. Rivers are typically not used because they are subject to drought and flooding, both of which may damage the system.

• Advantages: Can require the least total pipe length of closed-loop designs; can be less expensive than other closed-loop designs if body of water available.

• Disadvantages: Requires a large body of water and may restrict lake use (i.e., boat anchors).

Open-loop Systems

[15] Open-loop systems use local groundwater or surface water (i.e., lakes) as a direct heattransfer medium instead of the heat-transfer fluid described for the closed-loop systems. These systems are sometimes referred to specifically as "groundwater source heat pumps" to distinguish them from other GSHPs. Open-loop systems consist primarily of extraction wells, extraction and reinjection wells, or surface water systems. A variation on the extraction well system is the standing column well. This system reinjects the majority of the return water back into the source well, minimising the need for a reinjection well and the amount of surface discharge water. There are several special factors to consider in open-loop systems. **One major factor is water quality. In open-loop systems, the primary heat exchanger between the refrigerant and the groundwater is subject to fouling, corrosion, and blockage.** A second major factor is the adequacy of available water. The required flow rate

through the primary heat exchanger between the refrigerant and the groundwater is typically between 1.5 and 3.0 gallons per minute per system cooling ton [0.027 and 0.054 L/(s. kW)]. This can add up to a significant amount of water and can be affected by local water resource regulations. A third major factor is what to do with the discharge stream. The groundwater must either be re-injected into the ground by separate wells or discharged to a surface system such as a river or lake. Local codes and regulations may affect the feasibility of open-loop systems. Depending on the well configuration, open-loop systems can have the highest pumping load requirements of any of the ground-coupled configurations. In ideal conditions, however, an open-loop application can be the most economical type of ground-coupling system.

• Advantages: Simple design; lower drilling requirements than closed-loop designs; subject to better thermodynamic performance than closed-loop systems because well (s) are used to deliver groundwater at ground temperature rather than as a heat exchanger delivering heat-transfer fluid at temperatures other than ground temperature; typically lowest cost; can be combined with potable water supply well; low operating cost if water already pumped for other purposes, such as irrigation.

• Disadvantages: Subject to various local, state, and federal clean water and surface water codes and regulations; large water flow requirements; water availability may be limited or not always available; heat pump heat exchanger subject to suspended matter, corrosive agents, scaling, and bacterial contents; typically subject to highest pumping power requirements; pumping energy may be excessive if the pump is oversized or poorly controlled; may require well permits or be restricted for extraction; water disposal can limit or preclude some installations; high cost if reinjection well required.

[16] This type is characterised by the fact that the main heat carrier, ground water, flows freely in the underground, and acts as both a heat source/sink and as a medium to exchange heat with the solid earth. Main technical part of open systems is ground-water wells, to extract or inject water from/to water bearing layers in the underground ("aquifers"). In most cases, two wells are required ("doublette"), one to extract the groundwater, and one to re-inject it into the same aquifer it was produced from (Fig. 5).

[17] With open systems, a powerful heat source can be exploited at comparably low cost. On the other hand, groundwater wells require some maintenance, and open systems in general are confined to sites with suitable aquifers. The main requirements are:

• Sufficient permeability to allow production of the desired amount of groundwater with little draw down.

• Good groundwater chemistry e.g., low iron content to avoid problems with scaling, clogging and corrosion. Open systems tend to be used for larger installations. The most powerful GSHP system worldwide uses groundwater wells to supply ca. 10 MW of heat and cold to a hotel and offices.

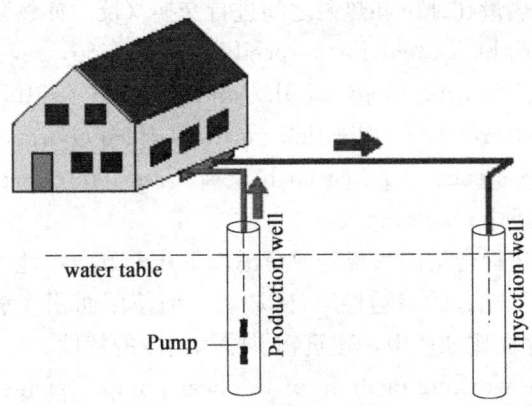

Fig. 5 Groundwater heat pump doublette

II. Words and Expressions

extraction	n.	加湿；浸取；提取；萃取
evaporator	n.	蒸发器
barrier	n.	屏障；界线
parameters	n.	参数；参量
parallel	n.	平行的
direct expansion		直接膨胀；直接蒸发
the refrigeration cycle	n.	制冷循环
horizontal-loop	n.	水平螺旋管
antifreeze	n.	防冻剂
borehole	n.	钻孔；挖孔
convective	adj.	对流的
adequate	adj.	充足的；适当的；胜任的
fouling	n.	污染；污垢
corrosion	n.	腐蚀；腐蚀产生的物质
blockage	n.	堵塞，堵塞
irrigation	n.	灌溉；冲洗
bacterial	adj.	细菌的

III. Notations

1. Closed systems: Heat exchangers are located in the underground (either in a horizontal, vertical or oblique fashion), and a heat carrier medium is circulated within the heat exchangers, transporting heat from the ground to the heat pump (or vice versa).

are located in 位于，放置于

闭式系统：换热器安放在地下（以水平的，垂直的，或者倾斜的形式），热量携带载

体在换热器内部循环，热量在地下和热泵之间进行交换（反之亦然）

2. To choose the right system for a specific installation, several factors have to be considered: geology and hydrogeology of the underground (sufficient permeability is a must for open systems), area and utilisation on the surface (horizontal closed systems require a certain area), existence of potential heat sources like mines, and the heating and cooling characteristics of the building (s)

为一个特定的装置选择合适的系统需要考虑以下几个方面：地下的地质和水文地质条件（一个开放的系统需要充分的渗透性），区域和表面使用面积（水平封闭系统需要一定的区域），潜在热源的存在比如矿山，建筑物供暖和制冷的特性。

3. In this case, the working medium of the heat pump (refrigerant) is circulating directly through the ground heat collector pipes (in other words, the heat pump evaporator is extended into the ground).

In this case 在这种情况下

在这种情况下，热泵系统的工作介质（制冷剂）直接在地面的收集管里循环（换句话说热泵的蒸发器直接延伸到地面）。

4. As noted with horizontal systems, slinky systems are also generally associated with lower tonnage systems where land area requirements are not a limiting factor.

are associated with 与……有关系；与……相关联

正如水平系统，螺旋系统也常常与较少的排水系统土地面积需求没有限制相关联

5. If more than one borehole heat exchanger is required, the pipes should be connected in such a way that equal distribution of flow in the different channels is secured.

如果系统需要多于一个钻孔换热器，管子的连接方式应使不同的管道内流量均等。

6. Submerged-loop applications require some special considerations, and it is best to discuss these directly with an engineer experienced in the design applications.

水下螺旋管的应用需要特别的考虑，最好直接与经验丰富的设计工程师讨论。

7. One major factor is water quality. In open-loop systems, the primary heat exchanger between the refrigerant and the groundwater is subject to fouling, corrosion, and blockage.

is subject to 常遭受……

其中一个主要因素是水的质量，在开式循环系统里，在制冷剂和地下水之间的主要换热器常遭受水污染，腐蚀和堵塞。

IV. Exercises

1. Translate the following sentences into Chinese.

(1) For the ground heat collectors with dense pipe pattern, usually the top earth layer is removed completely, the pipes are laid, and the soil is distributed back over the pipes.

(2) The main thermal recharge for all horizontal systems is provided for mainly by the solar radiation to the earth's surface. It is important not to cover the surface above the ground heat collector, or to operate it as a heat store, if it has to be located e. g., under a building.

(3) The spiral loop consists of pipe unrolled in circular loops in trenches and the horizontal configuration. Another variation of the spiral-loop system involves placing the loops upright in narrow vertical trenches.

(4) Requires more total pipe length than other ground coupled designs; relatively large ground area required; ground temperature subject to seasonal variance; larger pumping energy requirements than other horizontal loops defined above; backfilling the trench can be difficult with certain soil types and the pipe system could be damaged during backfill process.

(5) Coaxial (concentric) pipes, either in a very simple way with two straight pipes of different diameter, or in complex configurations heat exchangers in Europe.

(6) A third major factor is what to do with the discharge stream. The groundwater must either be re-injected into the ground by separate wells or discharged to a surface system such as a river or lake.

(7) This type is characterized by the fact that the main heat carrier, ground water, flows freely in the underground, and acts as both a heat source/sink and as a medium to exchange heat with the solid earth.

(8) Good groundwater chemistry e. g. , low iron content to avoid problems with scaling, clogging and corrosion. Open systems tend to be used for larger installations. The most powerful GSHP system worldwide uses groundwater wells to supply ca. 10 MW of heat and cold to a hotel and offices.

2. Translate the following sentences into English.

(1) 单U形管地热换热器的单位管长换热量比双U形管高，但是单U形管地热换热器的单位孔深换热量比双U形管低；经过分析知道，之所以单U形管地热换热器的单位管长换热量比双U形管高，是因为双U形管地热换热器各个管子之间的热短路现象所致。

(2) 冬季，当机组在制热模式时，就从土壤中吸收热量，通过压缩机和热交换器把大地的热量集中，并以较高的温度释放到室内。夏季，当机组在制冷模式时，就从土壤中提取冷量，通过压缩机和热交换器把大地的冷量集中并入室内，同时将室内的热量排放到土壤中，达到空调的目的。

(3) 土壤源热泵的污染物排放，与空气源热泵相比，相当于减少40%以上；与电采暖相比，相当于减少70 %以上；如果结合其他节能措施，节能减排效果会更明显。

Reading Material

A Brief History of Air Conditioning

[1] The term air conditioning most commonly refers to the cooling and dehumidification of indoor air for thermal comfort. In a broader sense, the term can refer to any form of cooling, heating, ventilation or disinfection that modifies the condition of air. An air conditioner (AC or A/C in North American English, aircon in British and Australian English) is an appliance, system, or mechanism designed to extract heat from an area, typically

using a refrigeration cycle but sometimes using evaporation, most commonly for comfort cooling in buildings and transportation vehicles.

[2] The concept of air conditioning is known to have been applied in Ancient Rome, where aqueduct water was circulated through the walls of certain houses to cool them. Similar techniques in medieval Persia involved the use of cisterns and wind towers to cool buildings during the hot season. Modern air conditioning emerged from advances in chemistry during the 19th Century, and the first large-scale electrical air conditioning was invented and used in 1902 by Willis Haviland Carrier.

[3] While moving heat via machinery to provide air conditioning is a relatively modern invention, the cooling of buildings is not. The ancient Romans were known to circulate aqueduct water through the walls of certain houses to cool them. As this sort of water usage was expensive, generally only the wealthy could afford such a luxury.

[4] Medieval Persia had buildings that used cisterns and wind towers to cool buildings during the hot season: cisterns (large open pools in a central courtyards, not underground tanks) collected rain water; wind towers had windows that could catch wind and internal vanes to direct the airflow down into the building, usually over the cistern and out through a downwind cooling tower. Cistern water evaporated, cooling the air in the building.

[5] In 1820, British scientist and inventor Michael Faraday discovered that compressing and liquefying ammonia could chill air when the liquefied ammonia was allowed to evaporate. In 1842, Florida physician Dr. John Gorrie used compressor technology to create ice, which he used to cool air for his patients in his hospital in Apalachicola, Florida. He hoped eventually to use his ice-making machine to regulate the temperature of buildings. He even envisioned centralized air conditioning that could cool entire cities. Though his prototype leaked and performed irregularly, Gorrie was granted a patent in 1851 for his ice-making machine. His hopes for its success vanished soon afterwards when his chief financial backer died; Gorrie did not get the money he needed to develop the machine. According to his biographer Vivian M. Sherlock, he blamed the "Ice King", Frederic Tudor, for his failure, suspecting that Tudor had launched a smear campaign against his invention. Dr. Gorrie died impoverished in 1855 and the idea of air conditioning faded away for 50 years.

[6] Early commercial applications of air conditioning were manufactured to cool air for industrial processing rather than personal comfort. In 1902 the first modern electrical air conditioning was invented by Willis Haviland Carrier. Designed to improve manufacturing process control in a printing plant, his invention controlled not only temperature but also humidity. The low heat and humidity were to help maintain consistent paper dimensions and ink alignment. Later Carrier's technology was applied to increase productivity in the workplace, and The Carrier Air Conditioning Company of America was formed to meet rising demand. Over time air conditioning came to be used to improve comfort in homes and automobiles. Residential sales expanded dramatically in the 1950s.

[7] In 1906, Stuart W. Cramer of Charlotte, North Carolina, USA, was exploring

ways to add moisture to the air in his textile mill. Cramer coined the term "air conditioning," using it in a patent claim he filed that year as an analogue to "water conditioning", then a well-known process for making textiles easier to process. He combined moisture with ventilation to "condition" and change the air in the factories, controlling the humidity so necessary in textile plants. Willis Carrier adopted the term and incorporated it into the name of his company. This evaporation of water in air, to provide a cooling effect, is now known as evaporative cooling.

[8] The first air conditioners and refrigerators employed toxic or flammable gases like ammonia, methyl chloride, and propane which could result in fatal accidents when they leaked. Thomas Midgley, Jr. created the first chlorofluorocarbon gas, Freon, in 1928. The refrigerant was much safer for humans but was later found to be harmful to the atmosphere's ozone layer. "Freon" is a trade name of Dupont for any Chlorofluorocarbon (CFC), Hydrogenated CFC (HCFC), or Hydrofluorocarbon (HFC) refrigerant, the name of each including a number indicating molecular composition (R11, R12, R22, R134). The blend most used in direct-expansion comfort cooling is an HCFC known as R22. It is to be phased out for use in new equipment by 2010 and completely discontinued by 2020. R-11 and R-12 are no longer manufactured in the US, the only source for purchase being the cleaned and purified gas recovered from other air conditioner systems. Several non-ozone depleting refrigerants have been developed as alternatives, including R410A, known by the brand name "Puron".

[9] Innovation in air conditioning technologies continue, with much recent emphasis placed on energy efficiency and for improving indoor air quality. As an alternative to high global warming refrigerants, such as R-134a in cars' and R22, R410a in residential air conditioning, natural alternatives like CO_2 (R744) have been proposed.

A Historical Perspective of Ventilation

[1] Most of us understand that the purpose of ventilation is to dilute indoor contaminants. (In this context internal gains can sometimes be a "contaminant" and ventilation for cooling is then an important consideration.) As engineers, we need to determine how much air is necessary to reduce concentrations of these contaminants to desired levels. Such an approach assumes some sort of scientific underpinnings.

[2] The science of ventilation and indoor air quality dates to the 17th century with work credited to Mayow. However, concerns about "bad air" go back much further in history. One of the earliest reasons to ventilate the indoor environment—and one that is still with us today—is to remove the products of combustion used for heat, light, or cooking. Early man undoubtedly learned quickly that if he brought fire in, he needed to get the smoke out. The specific driver for ventilation has changed over time, but has usually been associated with a particular set of pollutant sources that are causing health or comfort problems. Historically, these sources have been heat, combustion, people and their activities,

and the buildings themselves.

[3] Early living quarters offered a rudimentary level of protection from the elements, roaming creatures and foes. Caves, overhangs, and natural materials (e. g. , trees, stone, clay, animal skins) were used to build shelters. These shelters had sturdy doors and small openings to the outside to prevent any unwanted creatures or enemies from entering. This meant that homes were often smoky and that it was difficult to intentionally bring in significant outdoor air to dilute any indoor pollutants. Sanitary conditions were often poor, which resulted in both direct contamination as well as a growth medium for microorganisms.

[4] Archeological records provide several examples of how houses were built to accommodate ventilation and improve indoor air quality. Various approaches were developed to deal with the use of fires inside dwellings. In 4000 - 5000 B. C. , the Banpo villagers in China incorporated chimneys into their homes.

[5] The Romans put a vent hole in the middle of their houses' flat roofs to vent smoke out of the living quarters. The basket weaver's pit houses found in Mesa Verde National Park, circa 750 A. D. , use this same approach. Teepees, with their vent holes at the top and openings around the bottom were designed to accommodate fires and ventilation. The teepee doors could be positioned to control airflow.

[6] The Romans developed the hypocaust heating system for heating larger buildings. The hypocaust was a pre-cursor to the heating and ventilation system integrated into Britain's parliament buildings in the late 1800s. Outside air enters the hypocaust and gains heat from a fire and then is moved through an underfloor series of channels, up through channels in the walls and is vented to the outdoors. The systems built in the late 1880s use a similar approach but also supply outside air to the building. Outside ventilation air is pulled over steam pipes in a heating chamber and ducted into the building. Exhaust openings in the upper reaches of the building provide a stack effect to pull air through the building.

[7] In 1631, after finding that indoor conditions were causing health problems, King Charles I, in what may have been the first ventilation code, decreed that the ceilings in houses must be 10 ft (3 m) high or greater and that windows must be higher than their width to allow for ventilation. These improvements were slowly implemented into the British building stock. Implementation was hastened only when the great London fire of 1666 destroyed many of the inadequate houses and made way for construction of larger, better ventilated houses with chimneys and large windows. This trend towards better air quality, however, was suddenly thwarted and possibly reversed when citizens decided to board up their windows to avoid the chimney and window taxes of the early 1700s.

A Modern Perspective of Ventilation

[1] A look at past ventilation and indoor air quality issues provides insight into today's issues that might otherwise be overlooked. Ventilation is about health and most

regulations, recommendations and requirements have been driven by such acceptability concerns. Not surprisingly, it is ASHRAE's position to consider health impacts when setting criteria for indoor environments.

[2] Because of the effects it has on health, comfort, and serviceability, indoor air quality in our homes is an increasing concern to many people. According to the American Lung Association, elements within our homes have been increasingly recognized as threats to our respiratory health. The Environmental Protection Agency lists poor indoor air quality as the fourth-largest environmental threat to our country. Asthma is the leading serious chronic illness of children in the U.S. Moisture-related construction defects and damages are on the increase in new houses. Improved residential ventilation can reduce many of these indoor air quality problems.

[3] While increasing ventilation rates is often the first line of defense, source control always has been recognized as the *best* line of defense. The history of combustion moving from open and dirty combustion to sealed, clean combustion is an example. ASHRAE's position follows this trend by stating that source control is the most effective and preferred method of providing good indoor air quality in most cases.

[4] By the mid-20th century, society made major strides in controlling the sources of poor indoor air quality. Combustion appliances were becoming better. Many of the key infectious diseases had been eliminated as a major concern due to improved sanitation and hygiene as well as advances in the medical field (and a decrease in occupant densities). The best thinking of the time was that source control had been sufficiently successful that the biggest demand left for ventilation was to control the irreducible emissions of human bioeffluents.

[5] Unless they are infectious and when proper sanitation is practiced, human bioeffluents are not health hazards. These bioeffluents do, however, produce odors that can be unacceptable. The emission and acceptability of human bioeffluents have been well studied. Since the middle of the 20th century, the general assumption has been that if one ventilates to control human odors, there will be enough ventilation to control the health effects from other contaminants. Implicit in this approach is that sufficient care has been taken to reduce emissions of any contaminant below the level of concern. Nevertheless at a low enough ventilation rates there will always be indoor air quality problems.

Lesson 12 Building Automation Systems (建筑自动化系统)

I. Text

[1] The building automation system (BAS) has become the accepted technology used in controlling HVAC and other systems, in most new commercial and institutional buildings (Fig. 1). Existing buildings can be retrofitted with BASs, which has been shown to provide economically beneficial improvements in energy efficiency and occupant comfort.

[2] Although most BASs are designed primarily for HVAC control, many incorporate additional functions, such as lighting control, computerized maintenance scheduling, life-safety functions (such as smoke control) and access (security) control.

[3] Building automation systems, which are present in more than half of all buildings in the U.S. larger than 100,000 square feet, save an average of about 10% of overall building energy consumption. For older or poorly maintained buildings, the savings can be even greater. In addition to saving energy, these systems may also reduce the costs of overall building maintenance.

[4] Unfortunately, many building automation systems save less energy than they

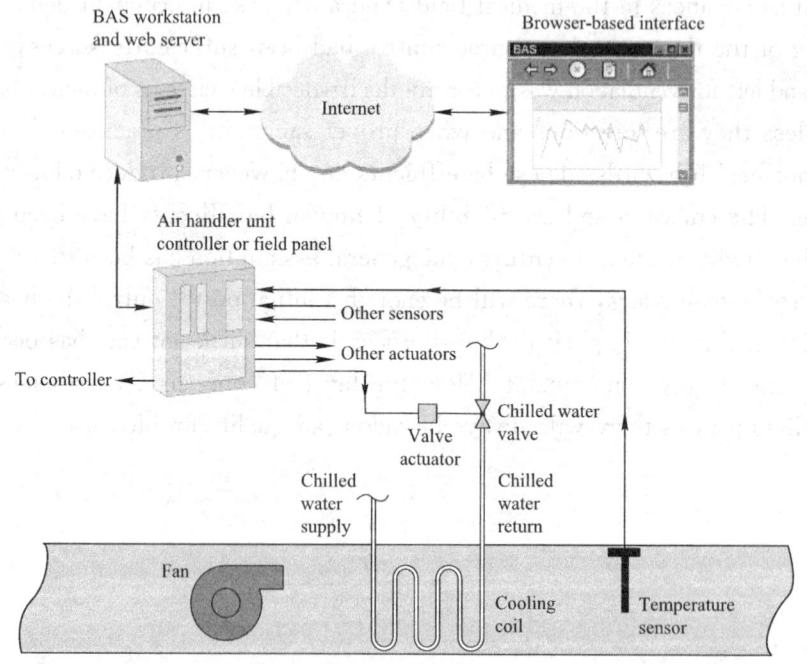

Fig. 1 How building automation systems fit together

A building automation system (BAS) consists of sensors, controllers, actuators and software.
An operator interfaces with the system via a central workstation or web browser.

could if set up optimally. In one detailed study of 11 buildings in New England with BASs, five of the buildings were found to be underachievers, producing less than 55% of expected savings. One site produced no savings at all.

[5] **To improve the likelihood that your BAS will achieve the expected benefits, you should take advantage of advanced control strategies that use the computer-processing power of a BAS and adopt a comprehensive approach to quality control, known as commissioning.** This process is now required for some buildings, such as public institutions and buildings certified by LEED (Leadership in Energy and Environmental Design). **Commissioning includes reviews and inspections throughout the design and construction process as well as rigorous performance tests that move the system through its sequences of operation before the building is occupied.** Recommissioning—when building operators use trending and energy consumption data to periodically verify, document and improve a building's operation—can be conducted throughout the life of the building.

What are the Options?

[6] There are many decisions that designers face when specifying a BAS, including:
- What control strategies to implement;
- Extent of the control to be provided by the BAS;
- Type of communications protocols to utilize; and
- Use of a web browser interface and Internet communications.

Energy-saving Control Strategies

[7] Here is a list of the most common strategies that building automation systems employ in order to use energy more efficiently:

[8] Scheduling. Scheduling is the practice of turning equipment on or off, depending on time of day, day of the week, day type or other variables, such as outside air conditions. Improving equipment schedules is one of the most common and effective measures for saving energy in commercial buildings. A feature called "Optimum Start", offered by all BAS manufacturers, can increase energy savings by automatically starting a system no earlier than necessary based on daily variations in the weather.

[9] Lockouts. Lockouts ensure that equipment does not come on unless necessary. They protect against nuances in the programming of the control system that may inadvertently cause equipment to turn on. For example, a chiller and its associated pumps can be locked out according to a calendar date, when the outside air falls below a certain temperature or when building cooling requirements are below a minimum.

[10] Resets. HVAC systems typically use less energy when their operating parameters are adjusted to meet the building load. Because this load varies with the weather, a BAS can help equipment to operate at greater efficiency levels by automatically varying these operating parameters. The simplest approach is to use a proportional reset schedule based on outdoor temperature (Fig. 2). Although that method works reasonably well, a more effective method is to base resets directly on building loads (Fig. 3). Examples of building control parameters that can be reset include supply-air and discharge-air tempera-

ture for fan systems that use terminal reheat, hot-deck and cold-deck temperatures for multizone HVAC systems, and heating-water supply temperature.

[11] Direct digital control (DDC). Direct digital control is provided by a BAS that directly controls valves, dampers and other system components for building temperature control (as shown in Fig. 1). The advantage of this approach (as opposed to using older conventional pneumatic or electronic controls) is that a more advanced control algorithm called PID (or proportional-integral-derivative) can be implemented in the BAS's computer code. Due to the complexity of this algorithm, older pneumatic or electronic controls used only the proportional form of this control technique, which is known for its inability to reliably maintain the temperature setpoint. If applied properly, PID can both save energy and provide improved comfort. However, note that proper implementation of a PID control algorithm is a complex process that is best left to experienced professionals.

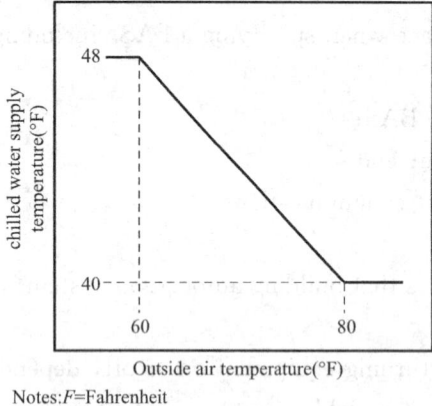

Notes: F=Fahrenheit

Fig. 2   Proportional reset schedule

As the outside air temperature decreases, the chilled water temperature is reset to a higher value.

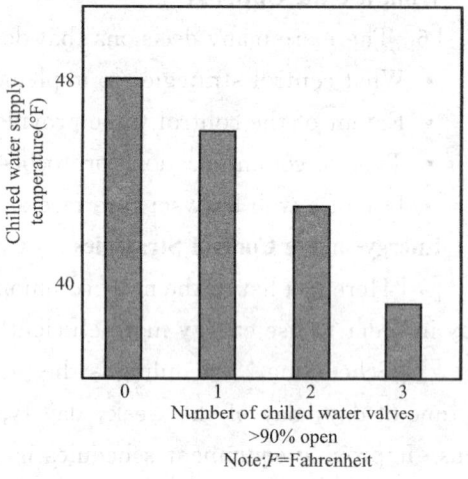

Note: F=Fahrenheit

Fig. 3   Direct load information reset

In this reset schedule, the cooling load is based on the number of chilled water valves that are greater than 90% open.

[12] Demand limiting. Because electrical demand charges can comprise 40% or more of a utility bill, many building automation systems can benefit from demand-limiting or load-shedding functions. For example, when the demand on a building meter or piece of equipment, such as a chiller, approaches a predetermined setpoint, the BAS does not allow the equipment to load up any further. In buildings with electric heat, electrical demand charges can be reduced if the heat is staged on in the morning over a several-hour period, starting with the coldest spaces first. Other demand-limiting strategies are expected to be developed as utilities implement time-of-day or real-time electrical pricing in their rate structures.

[13] Diagnostics. Building operators who use a BAS to monitor information—such as temperatures, air and water flows and pressures, and actuator positions—may use that data to determine whether equipment is operating incorrectly or inefficiently and troubleshoot

problems. A thorough job of building diagnostics typically requires the building operator to monitor more points than the minimal number needed to simply control a building, but a modern BAS gives users a good head start on a recommissioning or ongoing commissioning program. A modern BAS helped personnel from Texas A&M's Energy Systems Laboratory to cut energy bills at a state office building by 27%. The BAS helped to implement nighttime shutdowns and chart actual building temperatures to identify opportunities for temperature setback during unoccupied hours.

Extent of Control Provided by the BAS

[14] A major BAS design issue for both new and existing buildings is the extent of control provided by the BAS. This issue is handled differently for new buildings than for BAS retrofits in existing buildings:

[15] New buildings. HVAC equipment is often packaged with temperature controls provided by the equipment manufacturer. In many cases, these controls are a necessary part of the equipment and cannot be removed (for example, boiler and chiller operating controls, and safety limits). In this case, the BAS can still provide a certain amount of control that will improve the efficiency and comfort of the building. For example, a BAS usually provides chiller, pump and cooling tower staging for plants with multiple chillers. The BAS can also reset the chiller's operating parameters and monitor other chiller operating conditions. Note that boiler and chiller manufacturers sometimes provide their own "central plant" control systems. These are usually not recommended because they diminish the BAS's capability to perform energy-efficient control and provide diagnostics.

[16] Manufacturer-provided controls for other equipment, such as air handling units (AHUs), rooftop units and variable air volume (VAV) boxes, are usually optional. The advantage of using manufacturer-provided controls is that it reduces project costs and can often provide better control (because the controls are designed around the equipment). The disadvantage is that a communications protocol connection to this equipment will be needed to fully allow the BAS to perform energy-efficient supervisory control and provide diagnostics. Even with this communications connection, the manufacturer controls sometimes do not allow the BAS to adequately perform the necessary supervisory control. Because of this, many designs involve the field-installation of BAS controls on HVAC system components.

[17] Existing buildings. Existing buildings with older conventional controls, such as pneumatic and/or electronic controls, present a different design challenge. The existing hardware may limit the extent to which the BAS can provide system control. This challenge is due to the payback associated with the cost of replacing conventional controls versus the energy-savings benefit provided by the BAS. As a general rule, the use of a BAS to replace existing conventional controls for central equipment, such as boilers, chillers and AHUs, usually results in a short energy-savings payback (less than five years). On the other hand, replacing conventional controls (typically a pneumatic thermostat) on terminal HVAC equipment (typically VAV boxes or reheat coils) usually involves energy-savings

paybacks that can be in the 10 to 20-year range. **Note that the improved comfort of building occupants and enhanced ability to monitor, understand and diagnose malfunctions in the HVAC system often provide value above and beyond the energy savings that a BAS offers.**

[18] **In general, BASs that provide full DDC offer many more benefits over systems that use manufacturer-provided controls or depend on older, conventional controls.**

Communications Protocol

[19] Two major communications choices are available: proprietary and open (or standard).

[20] Proprietary communications protocols. Some BAS manufacturers use proprietary protocols that will communicate only with their own control equipment. These protocols may also be used on HVAC equipment manufacturer's controls that, with the explicit permission from a BAS manufacturer, were designed to communicate with the BAS. Proprietary systems may allow backward or forward compatibility with equipment generations of the same manufacturer, but they don't allow ready intercommunication with other brands of BASs. That is one reason why this type of system is rapidly disappearing from the marketplace. Because a proprietary system does not communicate with other systems, the user's options for expansion of the BAS are limited. Choices are also reduced for the purchase of new equipment, which limits the user's bargaining power. However, proprietary systems do offer the advantage of a single source of responsibility when there are problems.

[21] Open communications protocols. Open systems use communications protocols with publicly available documentation, which are therefore open for use by all BAS manufacturers. There are two major choices for open communications protocols. ASHRAE (the American Society of Heating, Refrigerating and Air-Conditioning Engineers) published a communications standard, known as BACnet, in 1995. The other major option is LonWorks technology, which was originally created by the Echelon Corp. (and which still controls many aspects of the technology). Most manufacturers of building controls have allied themselves with one or both of these standards, though there appears to be a growing preference for BACnet over LonWorks.

[22] There are several advantages to using an open communications protocol for a BAS. First, there is the assurance that equipment from multiple manufacturers will be able to interact. Using BACnet products that have been listed by BACnet Testing Laboratory ("BTL listed") ensures that they have been tested to confirm compliance with BACnet standards. Using equipment with open protocols also creates a competitive bidding environment for system additions and renovations, which helps to limit costs. This situation also helps keep manufacturers that have on-site equipment from becoming too comfortable and ensures a good level of service and response to problems.

[23] Another advantage is the containment of expenses associated with interfacing the BAS to mechanical equipment. For example, it is normally difficult to extend the features of a BAS with proprietary communications to monitor temperatures, pressures and flows of a new chiller. If all additions to a system are specified as open protocol, however, interfa-

cing becomes easier and less expensive.

[24] The use of open protocols also reduces the need to run multiple software packages on the BAS workstation or to utilize specialized interface equipment to communicate with devices using different protocols. The result is lower system costs and training expenses, fewer maintenance agreements and spare parts, and a single mode of system access.

Web Browser Interface and Internet Communications

[25] The introduction of web browser interfaces is the most important BAS development since the introduction of open communications protocols. Web browser capability is available via a software package, which usually runs on a dedicated web server, or may even be built into the highest-level controllers provided with a BAS. It allows a user to access and view the BAS through the Internet using a computer that is running web browser software. Users can take advantage of this capability to monitor and control the BASs in multiple facilities from a single computer (Fig. 4).

[26] Connecting a BAS to the Internet allows it to communicate with other computer applications such as online weather-forecasting services. The concept of enterprisewide management for facilities throughout the world is exciting, whether it concerns the management of HVAC control for building comfort, fire and physical safety, security, or buying power. Procurement of electricity in a deregulated world, for example, can become a real-time, dynamic activity facilitated by the BAS. Use of an Internet communications protocol called XML may also help to boost the use of the Internet for building control.

XML: An Emerging Standard

[27] Many of the technology companies involved in data exchange over the Internet have developed custom Internet applications using the Internet protocol called XML (Extensible Markup Language). XML has emerged as the standard protocol for data exchange in many business sectors and has gained attention in the field of building automation. XML is similar to HTML (Hypertext Markup Language), the language used to create the web pages that you can view using your web browser. The XML technology uses tags, much like HTML data tags, to record the relationships among the data elements in a file. The data in an XML file can associate a device, such as a controller, with numerous objects, such as points, messages, and alarms. A computer reading the file will be able to "understand" the physical capabilities of the objects and configure the system accordingly. By contrast, the same data written with HTML would associate a list with the controller, but it would not enable the computer to interpret the relationship between the controller and the items in the list.

[28] By supporting XML for building automation, manufacturers give their customers the flexibility to configure the system on their own, use a configuration package from another manufacturer, or use a third-party software package that supports XML as a file format, such as Microsoft Excel and Microsoft Access. Because Microsoft is freely distributing its XML software engine, it's much easier for manufacturers, software developers, or

users to create custom applications that read and write XML data, possibly even reading proprietary configuration data files and exporting them in standard XML format.

[29] Due to the popularity of the Internet and XML applications, new standards are emerging that simplify the use of XML communications in BASs. ASHRAE recently published an extension to the BACnet standard that defines how BACnet protocol information can be communicated using XML (the associated tools called "Web Services"). Concurrently, an organization known as CABA (the Continental Automated Building Association) has developed a building automation-oriented protocol called oBIX that also utilizes XML. As these new standards see wider use, more and more aspects of building automation communications will undoubtedly migrate to use of the Internet. Further, the use of XML may allow building automation systems to seamlessly communicate with business enterprise software, such as accounting and business scheduling (for example, Outlook) packages.

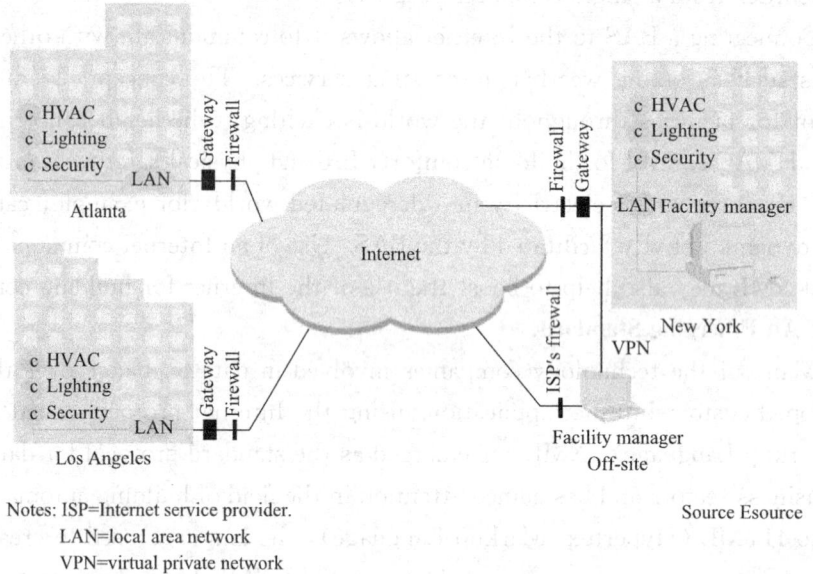

Fig. 4 How a web browser interface works

Controllers embedded in lighting, HVAC and security equipment communicate with each other via a local area network. Each building is then connected to the Internet through a gateway that is protected by a security firewall. Because these networked building systems offer remote control capabilities, facility managers can monitor and control their buildings from any location with a web connection. They can also manage multiple sites simultaneously or aggregate them for load control.

II. Words and Expressions

building automation system		楼宇自控系统
institutional buildings		公共建筑
retrofitted	adj.	式样翻新的,花样翻新的
incorporate	adj.	合并的,结社的,一体化的

	v.	合并
access (security) control		门警控制
underachiever	n.	未达标者
commissioning	n.	试车
communications protocols		通信协议
Energy-saving control strategies		节能控制模式
Scheduling	n.	行程安排，时序安排，例程
Optimum Start	n.	最优启动
lockouts	n.	停工（业主为抵制工人的要求而停工）
nuances	n.	细微差别
inadvertently	adv.	不注意地
reset	n.	复位
supply-air	n.	送风
discharge-air	n.	排风
terminal reheat		终端再热
hot-deck		热板
cold-deck		冷板
Direct digital control		直接数字控制
pneumatic		气动的
proportional-integral-derivative		比例积分微分
demand limiting	n.	需求限制
load-shedding	n.	用电限制
time-of-day		白天时间
real-time electrical pricing		实时电价，时段电价
diagnostics	n.	诊断学
troubleshoot		故障检修
recommissioning		重新试车
ongoing commissioning		正在进行的试车
central plant		重要车间，机房
variable air volume		变风量
field-installation		现场安装
proprietary	adj.	专有的，私有的
	n.	所有者，所有权
backward or forward compatibility	n.	向前或向后兼容性
open communications protocols		开放通信协议
BACnet		BACnet 协议
LonWorks		LonTalk 协议
competitive bidding		竞标，竞争出价
containment	n.	围堵政策，牵制政策

procurement	*n.*	获得，取得
XML (Extensible Markup Language)		可扩展标记语言
tag	*n.*	标签
proprietary configuration		专有配置
CABA (the Continental Automated Building Association)		北美大陆自动化建筑协会
automation-oriented		面向控制的
seamless	*adj.*	无缝合线的，无伤痕的
embedded	*adj.*	嵌入式的
local area network	*n.*	局域网
aggregate	*n.*	合计
	adj.	合计的
	v.	聚集，集合，合计

III. Notations

1. To improve the likelihood that your BAS will achieve the expected benefits, you should take advantage of advanced control strategies that use the computer-processing power of a BAS and adopt a comprehensive approach to quality control, known as commissioning.

known as commissioning 即 a comprehensive approach

为了提高 BAS 达到预期效益的可能性，应利用 BAS 具有基于计算机强大处理能力优势的先进控制策略，并采取全面质量控制方法即试车调试。

2. Commissioning includes reviews and inspections throughout the design and construction process as well as rigorous performance tests that move the system through its sequences of operation before the building is occupied.

that ... 修饰 tests

试车调试包括对整个设计和施工过程的检查和评价以及建筑投入运营之前对系统所进行的一系列严格的性能测试过程。

3. Note that the improved comfort of building occupants and enhanced ability to monitor, understand and diagnose malfunctions in the HVAC system often provide value above and beyond the energy savings that a BAS offers.

that ... 修饰 savings

注意：BAS 在改善居住者舒适感及提高暖通空调系统的故障监测诊断能力方面的价值远远超过它所能提供的节能价值。

4. In general, BASs that provide full DDC offer many more benefits over systems that use manufacturer-provided controls or depend on older, conventional controls.

offer 为谓语动词

一般来说，具备全面 DDC 功能的 BAS 系统能提供比基于设备制造商或较早的传统的控制方式的系统更多的优势。

IV. Exercises

1. Translate the following sentences into Chinese.

(1) Although most BASs are designed primarily for HVAC control, many incorporate additional functions, such as lighting control, computerized maintenance scheduling, life-safety functions (such as smoke control) and access (security) control.

(2) Recommissioning—when building operators use trending and energy consumption data to periodically verify, document and improve a building's operation—can be conducted throughout the life of the building.

(3) The advantage of this approach (as opposed to using older conventional pneumatic or electronic controls) is that a more advanced control algorithm called PID (or proportional-integral-derivative) can be implemented in the BAS's computer code.

(4) By supporting XML for building automation, manufacturers give their customers the flexibility to configure the system on their own, use a configuration package from another manufacturer, or use a third-party software package that supports XML as a file format, such as Microsoft Excel and Microsoft Access.

(5) Web browser capability is available via a software package, which usually runs on a dedicated web server, or may even be built into the highest-level controllers provided with a BAS.

2. Translate the following sentences into English.

(1) 建筑设备自动化系统（BAS）是智能建筑中的重要组成部分。BAS系统的主要功能是对智能建筑中的各种设备实行综合自动化管理，以达到舒适、安全、可靠、经济、节能的目的，为用户提供良好的工作和生活环境，并保证各项设备处于最佳运行状态。

(2) 建筑设备自动化系统主要包括暖通空调监控系统、建筑给水排水监控系统、建筑供配电监控系统、照明监控系统、交通监控系统、消防与安全防范系统、建筑设备自动化系统集成等。

(3) 建筑设备自动化系统是智能建筑弱电系统工程中较为复杂的系统之一，现将该系统的设计要点介绍如下。

Lesson 13　Particle Image Velocimetry （粒子图像测速）

I. Text

[1] Particle image velocimetry (PIV) is an optical method of fluid visualization. It is used to obtain instantaneous velocity measurements and related properties in fluids. The fluid is seeded with tracer particles which, for the purposes of PIV, are generally assumed to faithfully follow the flow dynamics. **It is the motion of these seeding particles that is used to calculate velocity information of the flow being studied.** Other techniques used to measure flows are Laser Doppler velocimetry and Hot-wire anemometry. The main difference between PIV and those techniques is that PIV produces two dimensional vector fields, while the other techniques measure the velocity at a point.

[2] **During PIV, the particle concentration is such that it is possible to identify individual particles in an image, but not with certainty to track it between images.** When the particle concentration is so low that it is possible to follow an individual particle it is called Particle tracking velocimetry, while Laser speckle velocimetry is used for cases where the particle concentration is so high that it is difficult to observe individual particles in an image.

[3] Typical PIV apparatus consists of a camera (normally a digital camera with a CCD chip in modern systems), a high power laser, for example a double-pulsed Nd: YAG laser or a copper vapor laser, an optical arrangement to convert the laser output light to a thin light sheet (normally using a cylindrical lens and a spherical lens), a synchronizer to act as an external trigger for control of the camera and laser, the seeding particles and the fluid under investigation. A fiber optic cable or liquid light guide often connects the laser to the lens setup. Fig. 1 shows a typical PIV setup.

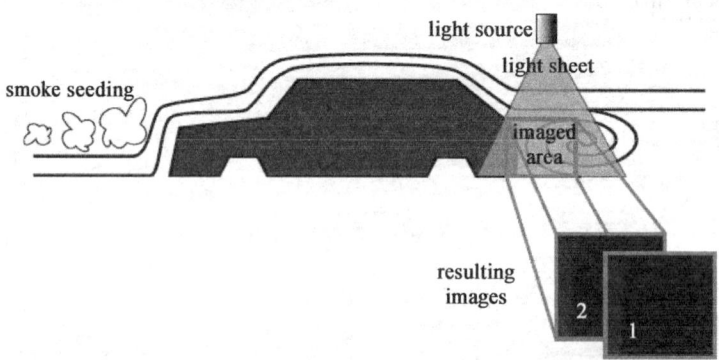

Fig. 1　Sketch of a typical setup for PIV ineasurements

Equipment and Aparatus

[4] The seeding particles are an inherently critical component of the PIV system. Depending on the fluid under investigation, the particles must be able to match the fluid properties reasonably well. Otherwise they will not follow the flow satisfactorily enough for the PIV analysis to be considered accurate. While the actual particle choice is dependent on the nature of the fluid, generally for macro PIV investigations they are glass beads, polystyrene, aluminum flakes or oil droplets (if the fluid under investigation is a gas). They must be inherently reflective, so that the laser sheet incident on the fluid flow will reflect off of the particles and be scattered towards the camera.

[5] The particles are typically of a diameter on the order of 10 to 100 micrometers. As for sizing, the particles should be small enough so that response time of the particles to the motion of the fluid is reasonably short to accurately follow the flow, yet large enough to scatter a significant quantity of the incident laser light. Due to the small size of the particles, the particle motion is dominated by stokes drag and settling or rising affects. Approximating the particles as spherical particles of very low Reynolds number, then the ability of the particles to follow the fluid's flow is directly proportional to the difference in density between the particles and the fluid and directly proportional to the square of the particles' diameters. The scattered light from the particles is dominated by Rayleigh scattering and so is also proportional to the square of the particles' diameters. Thus the particle size needs to be balanced to scatter enough light to accurately visualize all particles within the laser sheet plane but small enough to accurately follow the flow.

[6] The seeding mechanism needs to also be designed so as to seed the flow to a sufficient degree without overly disturbing the flow.

Camera

[7] To perform PIV analysis on the flow, two exposures of laser light are required upon the camera from the flow. Originally, with the inability of cameras to capture multiple frames at high speeds, both exposures were captured on the same frame and this single frame was used to determine the flow. A process called autocorrelation was used for this analysis. **However, as a result of auto-correlation the direction of the flow becomes unclear, as it is not clear which particle spots are from the first pulse and which are from the second pulse.** Faster digital cameras using CCD chips were developed since then that can capture two frames at high speed with a few hundred ns difference between the frames. This has allowed each exposure to be isolated on its own frame for more accurate cross-correlation analysis. The limitation of typical cameras is that this fast speed is limited to a pair of shots. This is because each pair of shots must be transferred to the computer before another pair of shots can be taken. Typical cameras can only take a pair of shots at a much slower speed. High speed CCD cameras are available but are much more expensive.

Laser and Optics

[8] For macro PIV setups, lasers are predominant due to their ability to produce high-

power light beams with short pulse durations. This yields short exposure times for each frame. Nd: YAG lasers, commonly used in PIV setups, emit primarily at 1064 nm wavelength and its harmonics (532, 266, etc.) For safety reasons, the laser emission is typically bandpass filtered to isolate the 532 nm harmonics (this is green light, the only harmonic able to be seen by the naked eye). A fiber optic cable or liquid light guide might be used to direct the laser light to the experimental setup. Fig. 2 shows the principal layout of a common PIV-laser.

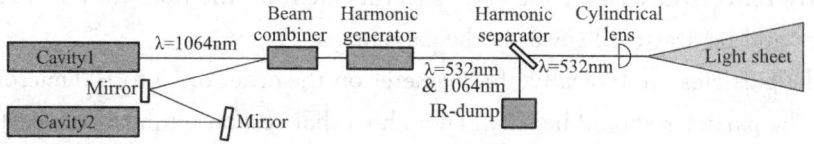

Fig. 2　Double cavity Nd: YAG PIV-laser

[9] The optics consists of a spherical lens and cylindrical lens combination. The cylindrical lens expands the laser into a plane while the spherical lens compresses the plane into a thin sheet. **This is critical as the PIV technique cannot generally measure motion normal to the laser sheet and so ideally this is eliminated by maintaining an entirely 2-dimensional laser sheet**. It should be noted though that the spherical lens cannot compress the laser sheet into an actual 2-dimensional plane. The minimum thickness is on the order of the wavelength of the laser light and occurs at a finite distance from the optics setup (the focal point of the spherical lens). This is the ideal location to place the analysis area of the experiment.

[10] The correct lens for the camera should also be selected to properly focus on and visualize the particles within the investigation area.

Synchronizer

[11] The synchronizer acts as an external trigger for both the camera (s) and the laser. While analogue systems in the form of a photosensor, rotating aperture and a light source have been used in the past, most systems in use today are digital. Controlled by a computer, the synchronizer can dictate the timing of each frame of the CCD camera's sequence in conjunction with the firing of the laser to within 1 ns precision. Thus the time between each pulse of the laser and the placement of the laser shot in reference to the camera's timing can be accurately controlled. Knowledge of this timing is critical as it is needed to determine the velocity of the fluid in the PIV analysis.

Analysis

[12] The frames are split into a large number of interrogation areas, or windows. It is then possible to calculate a displacement vector for each window with help of signal processing and autocorrelation or cross-correlation techniques. This is converted to a velocity using the time between laser shots and the physical size of each pixel on the camera. The size of the interrogation window should be chosen to have at least 6 particles per window on

average.

[13] The synchronizer controls the timing between image exposures and also permits image pairs to be acquired at various times along the flow. **For accurate PIV analysis, it is ideal that the region of the flow that is of interest should display an average particle displacement of about 8 pixels.** This is a compromise between a longer time spacing which would allow the particles to travel further between frames, making it harder to identify which interrogation window traveled to which point, and a shorter time spacing, which could make it overly difficult to identify any displacement within the flow.

[14] The scattered light from each particle should be in the region of 2 to 4 pixels across on the image. If too large an area is recorded, particle image size drops and peak locking might occur with loss of sub pixel precision. There are methods to overcome the peak-locking effect, but they require some additional work.

[15] If there is in house PIV expertise and time to develop a system, even though it is not trivial, it is possible to build a custom PIV system. Research grade PIV systems do, however, have high power lasers and high end camera specifications for being able to take measurements with the broadest spectrum of experiments required in research. If, for example, you want to spend less money, of course you get less resolution and lower framerates. There are also PIV analysis software available in the open source community. The results can have similar or even better quality compared to the expensive commercial PIV systems.

[16] Advantages:

• The method is to a large degree nonintrusive. The added tracers (if they are properly chosen) generally cause negligible distortion of the fluid flow.

• Optical measurement avoids the need for Pitot tubes, hotwires anemometers or other intrusive flow measurement probes. Additionally the method is capable of measuring an entire two-dimensional cross section (geometry) of the flow field simultaneously.

• High speed data processing allows the generation of large numbers of image pairs which, on a modern personal computer may be analysed in real time or at a later time. Thus a high quantity of near continuous information may be gained.

• Sub pixel displacement values allow a high degree of accuracy, since each vector is the statistical average for many particles within a particular tile. Displacement can typically be accurate down to 10% of one pixel on the image plane.

[17] Drawbacks:

• In some cases the particles will, due to their higher density, not perfectly follow the motion of the fluid (gas/liquid). If experiments are done e. g. in water, it is easily possible to find very cheap particles (e. g. plastic powder with a diameter of $\sim 60\mu m$) with the same density as water. If the density still does not fit, the density of the fluid can be tuned by increasing/decreasing its temperature. This leads to slight changes in the Reynolds number, so the fluid velocity or the size of the experimental object has to be changed to account for this.

- Particle image velocimetry methods will in general not be able to measure components along the z-axis (towards to/away from the camera). These components might not only be missed, they might also introduce interference in the data for the x/y-components. Some new methods also allow to measure three-dimensional flow though.
- Since the resulting velocity vectors are based on cross-correlating the intensity distributions over small areas of the flow, the resulting velocity field is a spatially averaged representation of the actual velocity field. This obviously has consequences for the accuracy of spatial derivatives of the velocity field, vorticity, and spatial correlation functions that are often derived from PIV velocity fields.
- Commercial research grade PIV systems include a Class IV laser and high resolution/speed digital camera that make the systems potentially unsafe and very expensive. Commercial systems are prohibitively expensive (around US＄100K).

[18] More Complex PIV Setups:

- Molecular tagging velocimetry, or MTV, uses molecule sized tags, which are often already a part of the flow. **Small molecules being much closer to the size and density of a flow minimize the error of particles not following the flow.**
- *Stereoscopic PIV* utilises two cameras with separate viewing angles to extract the z-axis displacement. Both cameras must be focused on the same spot in the flow and must be properly calibrated to have the same point in focus.
- *Dual Plane Stereoscopic PIV*

This is an expansion of stereoscopic PIV by adding a second plane of investigation directly offset from the first one. Four cameras are required for this analysis. The two planes of laser light are created by splitting the laser emission with a beam splitter into two beams. Each beam is then polarized orthogonally with respect to one another. Next, they are transmitted through a set of optics and used to illuminate one of the two planes simultaneously.

The four cameras are paired into groups of two. Each pair focuses on one of the laser sheets in the same manner as single-plane stereoscopic PIV. Each of the four cameras has a polarizing filter designed to only let pass the polarized scattered light from the respective planes of interest. This essentially creates a system by which two separate stereoscopic PIV analysis setups are run simultaneously with only a minimal separation distance between the planes of interest.

- *Micro PIV*

With the use of an epifluorescent microscope, microscopic flows can be analyzed. MicroPIV makes use of fluorescing particles that excite at a specific wavelength and emit at another wavelength. Laser light is reflected through a dichroic mirror, travels through an objective lens that focuses on the point of interest, and illuminates a regional volume. The emission from the particles, along with reflected laser light, shines back through the objective, the dichroic mirror and through an emission filter that blocks the laser light. Where PIV draws its 2-dimensional analysis properties from the planar nature of the laser sheet, microPIV utilizes the ability of the objective lens to focus on only one plane at a time, thus creating a 2-dimensional plane of viewable particles.

- *Holographic PIV*

Holographic PIV extracts the entirety of the motion of the particles in all planes. It does so by replacing a conventional light source with a laser. Particles struck by the laser will scatter light, and this will interfere with unscattered illumination, resulting in characteristic diffraction rings. By fitting this to Lorenz-Mie theory, the exact position of a particle in three dimensions may be determined with resolution of approximately 1 nm in-plane, and within 10 nm in the axial direction. By recording images of many particles flowing across the objective, velocity profiles may be reconstructed.

- *Scanning PIV*

By using a rotating mirror, a high-speed camera and correcting for geometric changes, PIV can be performed nearly instantly on a set of planes throughout the flow field. Fluid properties between the planes can then be interpolated. Thus, a quasi-volumetric analysis can be performed on a target volume. Scanning PIV can be performed in conjunction with the other 2-dimensional PIV methods described to approximate a 3-dimensional volumetric analysis.

- *Tomographic PIV*

A volumetric analysis technique, like holographic PIV, that utilizes four cameras oriented at different angles but focused on the same point to determine all of the acceleration properties of a fluid.

- Particle Tracking Velocimetry

By utilizing PIV analysis as an initial guess for a PTV analysis, more accurate analysis can be obtained than the spatial averaging of the cross-correlation method that typifies PIV analysis alone.

Applications

[19] PIV has been applied to a wide range of flow problems, varying from the flow over an aircraft wing in a wind tunnel to vortex formation in prosthetic heart valves. 3-Dimensional techniques have been sought to analyze turbulent flow and jets.

Rudimentary PIV algorithms based on cross-correlation can be implemented in a matter of hours, while more sophisticated algorithms may require a significant investment of time. Several open source implementations are available including URAPIV and mpiv (a Matlab Toolbox), PyPIV (an implementation in Python), JPIV (a Java implementation), OSIV and Gpiv (both implementations in C).

[20] Granular PIV: Velocity measurement in granular flows and avalanches

Particle image velocimetry (PIV) measurement technique is also introduced and used to measure the dynamics of the velocity distribution of free surface and unsteady flows of granular avalanches of non-transparent sand and quartz particles down channels and curved chutes merging into a horizontal plane from initiation to the run-out zone. Velocity distributions at the free surface are determined, and also at the bottom from below. These results can be applied to estimate impact pressures exerted by granular flows and avalanches on defence structures and infrastructures along the channel and in run-out zones.

PIV is used to measure the velocity field of the free surface and basal boundary in a granular flow and avalanche. This analysis is particularly designed for nontransparent fluids such as sand, gravel, quartz, or other granular materials that we can find in geophysical scenarios. This PIV system is called the "granular PIV". **The set-up of the granular PIV differs from the usual PIV in that the optical surface structure which is produced by illumination of the surface of the granular flow is already sufficient to detect the motion.** This means one does not need to add tracer particles in the bulk material.

II. Words and Expressions

Particle image velocimetry		粒子图像测速仪
fluid visualization		流动显示
seed	n.	种子,粒子
tracer particles		示踪物粒子
flow dynamics		流体动力学
Laser Doppler velocimetry		多普勒激光测速仪
Hot-wire anemometry		热线风速仪
vector fields		矢[向]量场
Laser speckle velocimetry		激光斑点测速仪
digital camera		数码相机
CCD chip		CCD 芯片
Nd:YAG laser		钕:钇-铝石榴子石激光器
copper vapor laser		铜蒸气激光器
cylindrical lens		柱面透镜
spherical lens		球面透镜
synchronizer	n.	同步装置,同步闪光装置
fiber optic cable		光纤导光臂
liquid light guide		液态导光臂
analog cameras		模拟相机
glass beads		玻璃珠
polystyrene	n.	聚苯乙烯
aluminum flakes		铝片
oil droplets		油滴
response time		响应时间

Stokes drag		史托克斯阻力
Rayleigh scattering		瑞利散射
exposures		曝光
frames	n.	帧，画面，框架
autocorrelation	n.	自相关
cross-correlation		互相关
bandpass filtered		带通滤波
displacement vector		位移矢量
signal processing		信号处理
peak-locking effect		锁峰效应
framerates		刷新率
Pitot tubes		皮托管
Intrusive	adj.	打扰的，插入的
pixel		像素
vorticity	n.	旋涡状态
correlation functions		互相关函数
tag		标签
hydroxy	n.	羟（基），氢氧基
hydroxyl tagging velocimetry		羟基示踪测速仪
stereoscopic	adj.	实体镜的，有立体感的
viewing angles		视角
Epifluorescent	adj.	寄生荧光的
dichroic mirror		分色镜，二向色镜
preprocessing		预加工，预处理
holographic	adj.	全息的
diffraction rings		［光］衍射环
topographic	adj.	地志的，地形学上的
granular	adj.	粒状的
prosthetic	adj.	词首增添字母［音节］的
run-out	n.	出局，超界
rudimentary	adj.	根本的，未发展的

III. Notations

1. It is the motion of these seeding particles that is used to calculate velocity information of the flow being studied.

注意 It is... that 从句的用法

正是利用这些粒子的运动来计算流场的速度信息。

2. During PIV, the particle concentration is such that it is possible to identify individual particles in an image, but not with certainty to track it between images.

but 之后省略了 it is

用 PIV 测速过程中，粒子的浓度应恰好可以识别一幅图像中的单个粒子，而不必一定要在多幅图像之间跟踪它们。

3. However, as a result of auto-correlation the direction of the flow becomes unclear, as it is not clear which particle spots are from the first pulse and which are from the second pulse.

注意第二个 as 引导原因状语从句

然而，由于自相关的原因，流动的方向变得不明确，因为尚不清楚那些光点是来此第一个脉冲的激光而那些是来此第二个脉冲的激光。

4. The cylindrical lens expands the laser into a plane while the spherical lens compresses the plane into a thin sheet. This is critical as the PIV technique cannot generally measure motion normal to the laser sheet and so ideally this is eliminated by maintaining an entirely 2-dimensional laser sheet.

注意 This 指整个前一句

柱面透镜将激光束扩展成平面光，而球面镜将此平面光压缩成一道薄片状的激光。这个过程是很关键的，因为 PIV 技术一般不能测量与薄片状光源垂直方向的运动。通过维持完全的二维激光片，垂直方向的运动可被理想地加以消除。

5. For accurate PIV analysis, it is ideal that the region of the flow that is of interest should display an average particle displacement of about 8 pixels.

注意第二个 that 为 region

对于准确的 PIV 分析而言，视窗大小能显示粒子约 8 个像素的平均位移是比较理想的。

6. Small molecules being much closer to the size and density of a flow minimize the error of particles not following the flow.

being much… 修饰主语 molecules，minimize 为谓语

越接近流场尺度和密度的小分子能最大限度地减小由于粒子不能跟随流动所造成的误差。

7. The set-up of the granular PIV differs from the usual PIV in that the optical surface structure which is produced by illumination of the surface of the granular flow is already sufficient to detect the motion.

in that… 引导原因状语从句

颗粒 PIV 装置不同于普通 PIV。由流动的颗粒表面产生的照度已足以用来检测运动。

IV. Exercises

1. Translate the following sentences into Chinese.

(1) Particle image velocimetry (PIV) is an optical method of fluid visualization. It is used to obtain instantaneous velocity measurements and related properties in fluids. The fluid is seeded with tracer particles which, for the purposes of PIV, are generally assumed to faithfully follow the flow dynamics.

(2) When the particle concentration is so low that it is possible to follow an individual particle it is called Particle tracking velocimetry, while Laser speckle velocimetry is used for cases where the particle concentration is so high that it is difficult to observe individual particles in an image.

(3) At these particle densities it was further noticed that it was easier to study the flows if they were split into many small 'interrogation' areas, which could be analyzed individually to generate one velocity for each area.

(4) They must be inherently reflective, so that the laser sheet incident on the fluid flow will reflect off of the particles and be scattered towards the camera.

(5) Knowledge of this timing is critical as it is needed to determine the velocity of the fluid in the PIV analysis.

(6) This is a compromise between a longer time spacing which would allow the particles to travel further between frames, making it harder to identify which interrogation window traveled to which point, and a shorter time spacing, which could make it overly difficult to identify any displacement within the flow.

(7) By fitting this to Lorenz-Mie theory, the exact position of a particle in three dimensions may be determined with resolution of approximately 1 nm in-plane, and within 10 nm in the axial direction. By recording images of many particles flowing across the objective, velocity profiles may be reconstructed.

(8) Particle image velocimetry (PIV) measurement technique is also introduced and used to measure the dynamics of the velocity distribution of free surface and unsteady flows of granular avalanches of non-transparent sand and quartz particles down channels and curved chutes merging into a horizontal plane from initiation to the run-out zone.

2. Translate the following sentences into English.

（1）要完成对流动的 PIV 技术分析，照相机需对来此流动的激光进行两次曝光。

（2）将各帧化为许多小区域或视窗，然后借助于信号处理和自相关或互相关技术，可计算每个视窗位移矢量。

（3）PIV 正是利用示踪粒子的散射特性来测量流体速度的。

Lesson 14 Solar Air Conditioning (太阳能空调)

I. Text

[1] Solar air conditioning refers to any air conditioning (cooling) system that uses solar power. This can be done through passive solar, solar thermal energy conversion and photovoltaic conversion (sun to electricity). Solar air conditioning will play an increasing role in zero energy and energy-plus buildings design.

Solar A/C Using Desiccants

[2] **Air can be passed over common, solid desiccants (like silica gel or zeolite) to draw moisture from the air to allow an efficient evaporative cooling cycle.** The desiccant is then regenerated by using solar thermal energy to dry it out, in a cost-effective, low-energy-consumption, continuously-repeating cycle. A photovoltaic system can power a low-energy air circulation fan, and a motor to slowly rotate a large disk filled with desiccant.

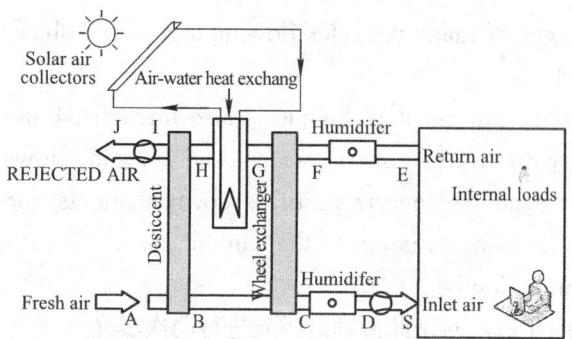
Fig. 1 Diagram of an installation of desiccant cooling

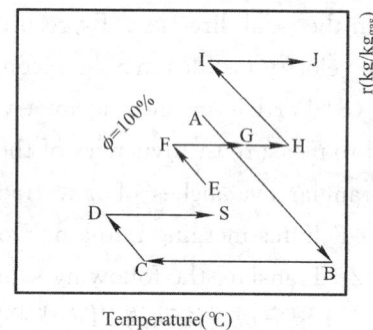
Fig. 2 Cycle's psychometric representation

[3] Energy recovery ventilation systems provide a controlled way of ventilating a home while minimizing energy loss. **Air is passed through an "enthalpy wheel" (often using silica gel) to reduce the cost of heating ventilated air in the winter by transferring heat from the warm inside air being exhausted to the fresh (but cold) supply air.** In the summer, the inside air cools the warmer incoming supply air to reduce ventilation cooling costs. **This low-energy fan-and-motor ventilation system can be cost-effectively powered by photovoltaics, with enhanced natural convection exhaust up a solar chimney - the downward incoming air flow would be forced convection (advection).**

[4] **A desiccant like calcium chloride can be mixed with water to create an attractive recirculating waterfall, that dehumidifies a room using solar thermal energy to regenerate the liquid, and a PV-powered low-rate water pump.**

[5] The potential for near-future exploitation of this type of innovative solar-powered

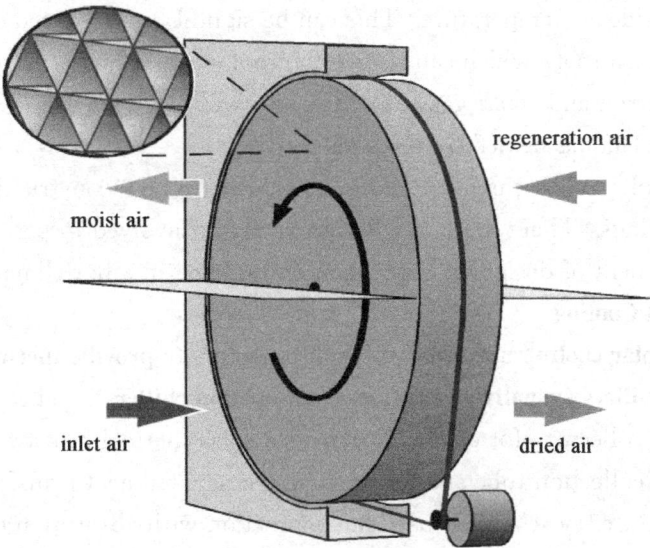

Fig. 3 Principle of a desiccant wheel
Lower (process) stream: inlet air-dried air.
Upper (regeneration) stream: regeneration air-moist air.

desiccant air conditioning technology is great.

[6] **Active solar cooling wherein solar thermal collectors provide input energy for a desiccant cooling system: A packed column air-liquid contactor has been studied in application to air dehumidification and regeneration in solar air conditioning with liquid desiccants.** A theoretical model has been developed to predict the performance of the device under various operating conditions. **Computer simulations based on the model are presented which indicate the practical range of air to liquid flux ratios and associated changes in air humidity and desiccant concentration.** An experimental apparatus has been constructed and experiments performed with Monoethylene Glycol (MEG) and Lithium Bromide (LiBr) as desiccants. MEG experiments have yielded inaccurate results and have pointed out some practical problems associated with the use of Glycols. LiBr experiments show very good agreement with the theoretical model. Preheating of the air is shown to greatly enhance desiccant regeneration. The packed column yields good results as a dehumidifier/regenerator, provided pressure drop can be reduced with the use of suitable packing.

Passive Solar Cooling

[7] In this type of cooling solar thermal energy is not used directly to create a cold environment or drive any direct cooling processes. Instead, passive solar building design aims at slowing the rate of heat transfer into a building in the summer, and improving the removal of unwanted heat. It involves a good understanding of the mechanisms of heat transfer: heat conduction, convective heat transfer, and thermal radiation, the latter primarily from the sun.

[8] For example, a sign of poor thermal design is an attic that gets hotter in summer

than the peak outside air temperature. This can be significantly reduced or eliminated with a cool roof or a green roof, which can reduce the roof surface temperature by 39℃ in summer. A radiant barrier and an air gap below the roof will block about 97% of downward radiation from roof cladding heated by the sun.

[9] Passive solar cooling is much easier to achieve in new construction than by adapting existing buildings. There are many design specifics involved in passive solar cooling. It is a primary element of designing a zero energy building in a hot climate.

Solar Thermal Cooling

[10] **Active solar cooling uses solar thermal collectors to provide thermal energy to drive thermally-driven chillers (usually adsorption or absorption chillers).** The Sopogy concentrating solar thermal collector, for example, provides solar thermal heat by concentrating the sun's energy on a collection tube and heating the recirculated heat transfer fluid within the system. The generated heat is then used in conjunction with absorption chillers to provide a renewable source of industrial cooling. (Fig. 4).

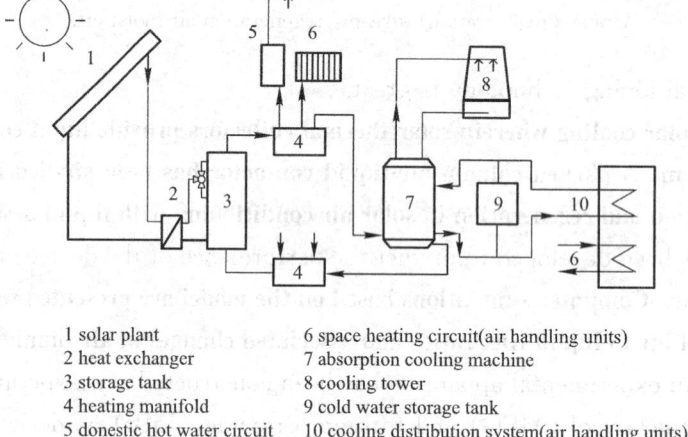

1 solar plant
2 heat exchanger
3 storage tank
4 heating manifold
5 domestic hot water circuit
6 space heating circuit(air handling units)
7 absorption cooling machine
8 cooling tower
9 cold water storage tank
10 cooling distribution system(air handling units)

Fig. 4 Principle of combines solar input with absorption chillers

[11] There are multiple alternatives to compressor-based chillers that can reduce energy consumption, with less noise and vibration. Solar thermal energy can be used to efficiently cool in the summer, and also heat domestic hot water and buildings in the winter. Single, double or triple iterative absorption cooling cycles are used in different solar-thermal-cooling system designs. The more cycles, the more efficient they are.

[12] In the late 1800s, the most common phase change refrigerant material for absorption cooling was a solution of ammonia and water. Today, the combination of lithium and bromide is also in common use. One end of the system of expansion/condensation pipes is heated, and the other end gets cold enough to make ice. Originally, natural gas was used as a heat source in the late 1800s. Today, propane is used in recreational vehicle absorption chiller refrigerators. Innovative hot water solar thermal energy collectors can also be used

as the modern "free energy" heat source.

[13] Efficient absorption chillers require water of at least 88℃. Common, inexpensive flat-plate solar thermal collectors only produce about 71℃ water, but several successful commercial projects in the US, Asia and Europe have shown that flat plate solar collectors specially developed for temperatures over 93.3℃ (featuring double glazing, increased backside insulation, etc.) can be effective and cost efficient. Evacuated-tube solar panels can be used as well. Concentrating solar collectors required for absorption chillers are less effective in hot humid, cloudy environments, especially where the overnight low temperature and relative humidity are uncomfortably high. Where water can be heated well above 88℃, it can be stored and used when the sun is not shining.

Photovoltaic (PV) Solar Cooling

[14] **Photovoltaics can provide the power for any type of electrically powered cooling be it conventional compressor-based or adsorption/absorption-based, though the most common implementation is with compressors which is the least efficient form of electrical cooling methods.**

[15] For small residential and small commercial cooling (less than 5 MWh/yr) PV-powered cooling has been the most frequently implemented solar cooling technology (Fig. 5). **The reason for this is debated, but commonly suggested reasons include incentive structuring, lack of residential-sized equipment for other solar-cooling technologies, the advent of more efficient electrical coolers, or ease of installation compared to other solar-cooling technologies** (like radiant cooling).

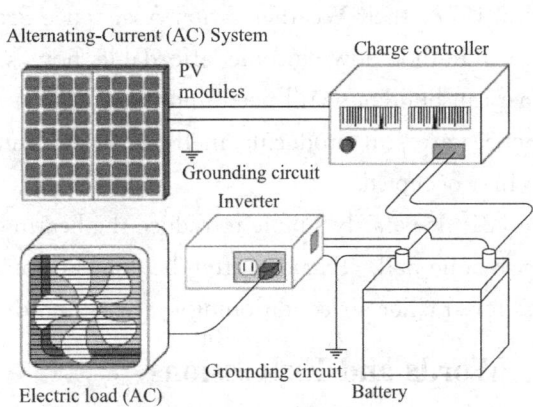

Fig. 5   PV-powered cooling

[16] **Since PV cooling's cost effectiveness depends largely on the cooling equipment and given the poor efficiencies in electrical cooling methods until recently it has not been cost effective without subsidies.** Pairing PV with 14 SEER and less coolers is the least efficient of all solar cooling methods. Using more efficient electrical cooling methods and allowing longer payback schedules is changing that scenario.

[17] A more efficient air conditioning system would require a smaller, less-expensive photovoltaic system. A high quality geothermal heat pump installation can have a SEER in the range of ±20. A 100,000 BTU SEER 20 air conditioner would require less than 5 kW while operating.

[18] There are new non-compressor-based electrical air conditioning systems with a SEER above 20 coming on the market. New versions of phase-change indirect evaporative coolers use nothing but a fan and a supply of water to cool buildings without adding extra interior humidity. In dry arid climates with relative humidity below 45% (about 40% of

the continental U. S.) indirect evaporative coolers can achieve a SEER above 20, and up to SEER 40. A 100,000 BTU indirect evaporative cooler would only need enough photovoltaic power for the circulation fan (plus a water supply).

[19] A less-expensive partial-power photovoltaic system can reduce (but not eliminate) the monthly amount of electricity purchased from the power grid for air conditioning (and other uses). With American state government subsidies of $2.50 to $5.00 USD per photovoltaic watt, the amortized cost of PV-generated electricity can be below $0.15 per kWh. This is currently cost effective in some areas where power company electricity is now $0.15 or more. **Excess PV power generated when air conditioning is not required can be sold back to the power grid in many locations, which can reduce (or eliminate) annual net electricity purchase requirement.**

[20] The key to solar air conditioning cost effectiveness is in lowering the cooling requirement for the building. Superior energy efficiency can be designed into new construction (or retrofitted to existing buildings). Since the U. S. Department of Energy was created in 1977, their Weatherization Assistance Program has reduced heating-and-cooling load on 5.5 million low-income affordable homes an average of 31%. A hundred million American buildings still need improved weatherization. Careless conventional construction practices are still producing inefficient new buildings that need weatherization when they are first occupied.

[21] It is fairly simple to reduce the heating-and-cooling requirement for new construction by one half. This can often be done at no additional net cost, since there are cost savings for smaller air conditioning systems and other benefits.

II. Words and Expressions

passive	*adj.*	被动的
photovoltaic (PV)		光电的
zero energy		零功率，零能耗
energy-plus buildings		节能大厦
Solar A/C		太阳能空调
desiccants	*n.*	干燥剂
	adj.	使干燥的，去湿的
silica gel	*n.*	硅胶
zeolite	*n.*	沸石
energy recovery ventilation		能量回收式通风
enthalpy wheel		潜热（焓）转轮
fan-and-motor		风扇电动机，风机马达
solar chimney		太阳能烟囱
advection	*n.*	水平对流
calcium chloride		氯化钙

dehumidify	vt.	除湿，使干燥
PV-powered	adj.	光伏供电的
low-rate		低速率
exploitation	n.	开发，开采
solar thermal collectors		太阳热能集热器
packed	adj.	紧凑的
air-liquid contactor		气液接触器
regeneration	n.	再生，重建
concentration	n.	浓缩，浓度
Monoethylene Glycol (MEG)		单乙烯-乙二醇
Lithium Bromide		溴化锂
aim at	v.	瞄准，针对
attic	n.	阁楼，顶楼
cool roof	n.	冷屋顶
green roof	n.	绿色屋顶
radiant barrier	adj.	发光的，辐射的
air gap	n.	间隙
roof cladding	n.	屋顶覆层
recirculate	v.	再通行，再流通
absorption chillers	n.	吸收冷却器
phase change refrigerant	n.	相变制冷剂
ammonia	n.	氨，氨水
propane	n.	丙烷
recreational	adj.	休养的，娱乐的
featuring double glazing		配置双层玻璃
evacuated-tube		退役管
intermittent	adj.	间歇的，断断续续的
parabolic trough solar collector		抛物线槽式太阳集热器
cost effective		有效成本
BTU (British Thermal Unit)		英国热量单位
seasonal energy efficiency ratio (SEER)		季节性能量效率比
seasonal solar tracker capability		季节性太阳追踪能力
rated	adj.	定价的，额定的
solar altitude		太阳高度
well over		溢出，超过
net metering	n.	净值测量（法）
utility companies		公网公司
geothermal heat pump		地源热泵
indirect evaporative coolers		间接蒸发冷却器

power grid　　　　　　　　　　　　　　　电网
amortize　　　　　　　　v.　　　　　　分期清偿
low-income affordable home　　　　　　低收入可承担家庭

III. Notations

1. Air can be passed over common, solid desiccants (like silica gel or zeolite) to draw moisture from the air to allow an efficient evaporative cooling cycle.

　　to draw... 修饰 solid desiccants, to allow.... 表示目的

　　空气可以通过共用固体干燥剂（如硅胶或分子筛）吸取空气中的水分蒸发，而产生高效的蒸发冷却循环。

2. Air is passed through an "enthalpy wheel" (often using silica gel) to reduce the cost of heating ventilated air in the winter by transferring heat from the warm inside air being exhausted to the fresh (but cold) supply air.

　　by transferring... 方式状语

　　空气通过一个"焓轮"（通常采用硅胶），将排风中的热量传递给新风以减少冬季热风采暖的成本。

3. This low-energy fan-and-motor ventilation system can be cost-effectively powered by photovoltaics, with enhanced natural convection exhaust up a solar chimney - the downward incoming air flow would be forced convection (advection).

　　with... 方式状语

　　借助太阳能烟囱所强化的自然对流——向下流入的空气被强制对流，这种低能耗机械通风系统可由太阳能光伏有效驱动。

4. A desiccant like calcium chloride can be mixed with water to create an attractive recirculating waterfall, that dehumidifies a room using solar thermal energy to regenerate the liquid, and a PV-powered low-rate water pump.

　　that... 结果状语

　　将像氯化钙这样的干燥剂与水混合可产生一个很引人注意的循环水幕，这样就可利用太阳能再生的液体及光伏驱动的低速水泵来使室内达到除湿的目的。

5. Active solar cooling where in solar thermal collectors provide input energy for a desiccant cooling system: A packed column air-liquid contactor has been studied in application to air dehumidification and regeneration in solar air conditioning with liquid desiccants.

　　注意 where in... 用法

　　主动式太阳能制冷即以太阳能集热器为其提供能量的干燥剂冷却系统：填充柱气液接触器已被研究应用于液体干燥剂太阳能空调中空气的除湿和液体干燥剂的再生。

6. Computer simulations based on the model are presented which indicate the practical range of air to liquid flux ratios and associated changes in air humidity and desiccant concentration.

　　which... 修饰 Computer simulations

基于该模型的计算机模拟已给出了气液流量比及空气湿度和干燥剂浓度相应变化的实用范围。

7. Active solar cooling uses solar thermal collectors to provide thermal energy to drive thermally-driven chillers (usually adsorption or absorption chillers).

注意 use sth. to do sth. 的用法

主动式太阳能冷却系统利用太阳能集热器所提供的热能来驱动热驱动制冷机（通常是吸附或吸收式制冷机）。

8. Photovoltaics can provide the power for any type of electrically powered cooling be it conventional compressor-based or adsorption/absorption-based, though the most common implementation is with compressors which is the least efficient form of electrical cooling methods.

注意 be it A or B means whether it is A or B。be it 是谓语提前的倒装用法，it 是形式主语。这里不指代具体某个词，某件事

光伏发电可为任何类型的电动冷却系统提供动力，无论是传统的压缩机型还是吸附/吸收型，尽管最常见的情况是以电冷却系统中最低效的压缩机方式实现的。

9. The reason for this is debated, but commonly suggested reasons include incentive structuring, lack of residential-sized equipment for other solar-cooling technologies, the advent of more efficient electrical coolers, or ease of installation compared to other solar-cooling technologies (like radiant cooling).

注意 reason for this 中的 this 指前一句所表述的事情

这样做的原因是值得讨论的，但通常建议的原因包括了激励政策，缺乏适合住宅大小的其他太阳能冷却技术设备及有效的电力冷却设备的出现，或相对于其他太阳能冷却技术（如辐射冷却）方式而言安装较为方便。

10. Since PV cooling's cost effectiveness depends largely on the cooling equipment and given the poor efficiencies in electrical cooling methods until recently it has not been cost effective without subsidies.

注意 given... 鉴于...，作介词用

由于光伏冷却的成本效益在很大程度上取决于冷却设备，再加上电制冷却方式的效率低，直到最近，如果没有补贴的话其效益还是很差的。

11. Excess PV power generated when air conditioning is not required can be sold back to the power grid in many locations, which can reduce (or eliminate) annual net electricity purchase requirement.

which... 表示前一句所表述的事情

在许多地方，当空调不用电时光伏所产生的过剩电量可卖给电网，这样就可以减少（或消除）年度从电网净购电的需求。

IV. Exercises

1. Translate the following sentences into Chinese.

(1) The desiccant is then regenerated by using solar thermal energy to dry it out, in a

cost-effective, low-energy-consumption, continuously-repeating cycle. A photovoltaic system can power a low-energy air circulation fan, and a motor to slowly rotate a large disk filled with desiccant.

(2) In this type of cooling solar thermal energy is not used directly to create a cold environment or drive any direct cooling processes. Instead, passive solar building design aims at slowing the rate of heat transfer into a building in the summer, and improving the removal of unwanted heat.

(3) Single, double or triple iterative absorption cooling cycles are used in different solar-thermal-cooling system designs. The more cycles, the more efficient they are.

(4) Concentrating solar collectors required for absorption chillers are less effective in hot humid, cloudy environments, especially where the overnight low temperature and relative humidity are uncomfortably high.

(5) Due to the advent of net metering allowed by utility companies, your photovoltaic system will produce enough energy in the course of the year to completely offset the cost of the electricity used to run air conditioning.

(6) New versions of phase-change indirect evaporative coolers use nothing but a fan and a supply of water to cool buildings without adding extra interior humidity.

(7) Zero energy buildings (with solar air conditioning, etc.) can help solve our current energy crisis, reduce the largest contributor to global warming, improve energy independence, and energy security, create millions of new green-collar worker jobs, and greatly increase gross domestic product, while eliminating unnecessary utility bills, and extending the useful life of our finite electric power generation facilities.

2. Translate the following sentences into English.

(1) 所谓太阳能制冷，就是利用太阳集热器为吸收式制冷机提供其发生器所需的热媒水。热媒水温越高，制冷机的COP也越高，这样太阳能空调系统的效率也越高。

(2) 与全玻璃真空管（all-glass tubular）集热器相比，金属吸热体真空管集热器具有承压能力强和耐热冲击（thermal shock）好两大优点，这些优点对于大多数主动太阳能系统来说是非常必要的。

(3) 利用太阳能为能源的空调系统，它的优势就在于越是太阳辐射强烈时，环境气温越高，人们的生活越需要空调，此时，太阳能空调的制冷效果越好。

Lesson 15　Introduction to Thermal Comfort（热舒适性简介）

I. Text

[1] **This section gives an overview of how human thermal comfort is affected by the body's physiology, and the external conditions surrounding the body.**

[2] Thermal comfort is defined as that condition of mind that expresses satisfaction with the thermal environment. Dissatisfaction may be caused by warm or cool discomfort of the body as a whole or may be caused by an unwanted cooling (or heating) of one particular part of the body.

[3] Due to individual differences, it is impossible to specify a thermal environment that will satisfy everybody. There will always be a percentage of dissatisfied occupants. But it is possible to specify environments (using indices later defined) predicted to be acceptable by a certain percentage of the occupants.

[4] To understand how to best control comfort, we must first understand the human body's method of maintaining heat balance.

Heat Production Within the Body

[5] The human body must maintain heat balance if it is to survive. If it generates more heat than is needed, it must lose heat to its surroundings or its temperature will rise and it will become ill and could die. Likewise, if it loses too much heat to maintain a constant temperature, its temperature will lower and it could die. When conditions surrounding the body are not ideal, it has adaptation mechanisms that help adjust the amount of heat loss. When the body has to take adaptive measures, it is known to be under thermal stress. Thermal stress equates to discomfort while a minimum of thermal stress provides comfort.

[6] The human body can be compared to a machine (in the engineering world) that converts fuel into energy for the purpose of doing work —— the more active the body, the more fuel that is consumed. The rate of heat production within the body is known as the metabolic rate (units=met= 360 Btu/hr) and includes all of the heat given off by all of the chemical reactions taking place in the body. Some examples of typical metabolic rates are as follows:

Activity	No. of Mets
Sleeping	0.7
Seated, quiet (office work)	1.0
Walking (3 mph)	2.6
Tennis, singles	4.6

[7] Like the machine, the conversion of fuel (oxidation of food) into work is not 100% efficient. **That energy which is not converted to do work is in the form of heat, and if not needed to maintain a constant body temperature, it is brought to surfaces by blood flow**

(**Fig. 1**), **then rejected to the body's surrounding environment.** This heat is rejected in two forms, sensible and latent heat transfer.

The Body's Thermoregulatory System

[8] The heat released within the body warms the blood which circulates to all body tissues, keeping them at the homeostatic temperature. The body temperature is a result of the balance between heat production and heat loss. The hypothalamus is the body's thermostat. Located in the brain, the hypothalamus continuously regulates the body's temperature, using the nervous system's pathways, to a constant setpoint of around 37.7℃.

Increase Heat Loss

[9] The body must be protected from excessively high temperatures. Heat loss from the skin surface occurs mostly from radiation or evaporation. As the body's temperature increases above what is desirable, the warm blood comes to the skin via dialated blood vessels and the capillary beds in the skin become flushed with the warm blood. The result is heat radiating from the skin surface.

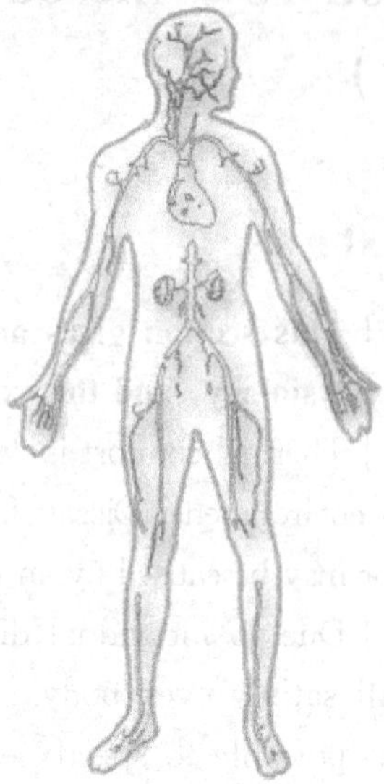

Fig. 1 There are three sensible heat loss mechanisms: Radiant loss to cooler surfaces (or gain from warmer surfaces); Convection loss to cooler air (or gain from warmer air) which is heated and rises; and dry respiration heat loss to cooler air that enters the lungs and is exhaled warmer. The latent heat loss mechanisms include: latent respiration heat loss; water diffusion through the skin; and the evaporation of sweat (skin wettedness).

[10] **Over the years, researchers of human comfort have established the variables that affect a human's thermal sensations and they have established the ranges of these variables within which the average person is comfortable.**

Six Primary Comfort Variables

[11] These "comfort" variables include air temperature, relative humidity, air motion, and mean radiant temperature (Fig. 2). The mean radiant temperature is the average temperature of all of the surfaces that surround the person in question. These four variables are called the "environmental variables" because they represent the environment surrounding the body.

[12] A second set of variables, called the "personal" variables, are controlled by the individual. The two are the clothing insulation value, termed the "clo" value, and the metabolism rate, with units of "met" as described above.

[13] Comfort conditions have been difficult to visualize. Most people, mistakenly, equate comfort with an air temperature range. **One hundred years ago, Willis Carrier, developed a method that allows us to visualize two of the variables — the combination of air temperature and relative humidity that exist in a space.** His tool is called the Psychrometric Chart (Fig. 3). Psychrometrics, which Willis Carrier imagined and developed, is the

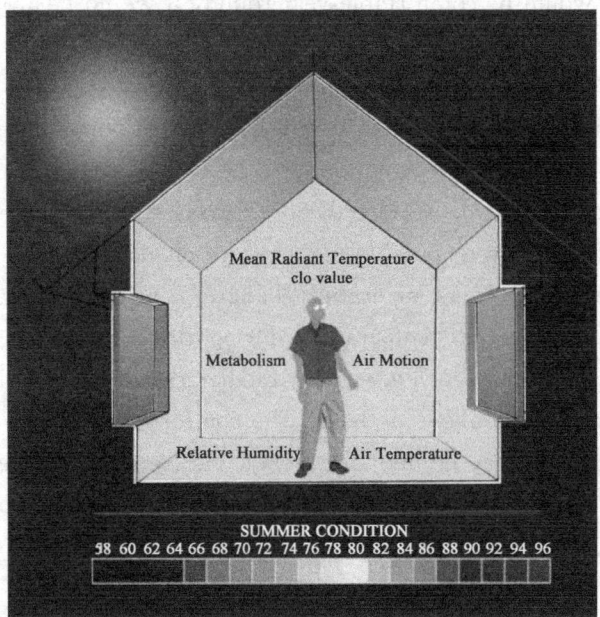

Fig. 2 Six Primary Comfort Variables

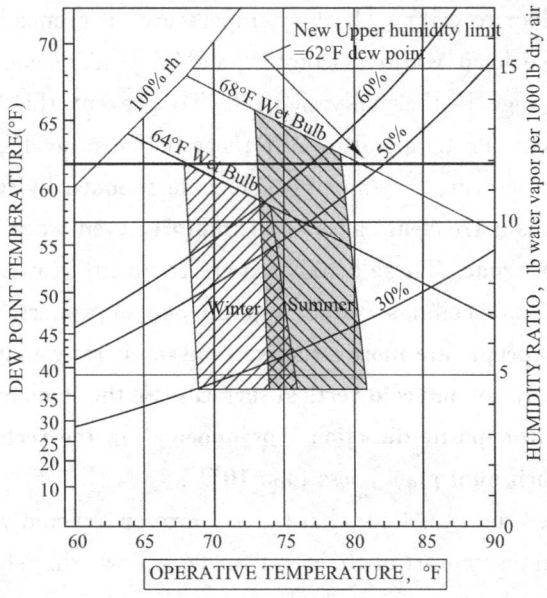

Fig. 3 Psychrometric Chart

study of air and water vapor mixtures, and is the scientific basis of the air conditioning industry. Below we see a Psychrometric chart that outlines the conditions at which most sedentary humans are comfortable; one envelope is for winter comfort and one for summer comfort. There is a difference because of the amount of clothing worn in the two different seasons. Any point located on the chart establishes the temperature (dry bulb) and the amount of water vapor in a unit quantity of air. If we were to select one temperature at which most humans would be comfortable year round, it would be around 23℃.

Other Parameters that Affect Comfort

Time

[14] During passive comfort control or when thermal mass is employed to achieve comfort conditions, temperature fluctuations will be experienced in the form of "drifts" or "ramps." The comfort standard (ASHRAE 55) discusses these "non-steady state" conditions. Generally, drifts refer to passive temperature changes of the enclosed space, and ramps refer to actively controlled temperature changes. The comfort standard specifies that the rate of change in operative temperature during drifts or ramps should not exceed 0.6℃/h. Drifts or ramps are allowed without further restriction provided that the operative temperature does not go above or below the comfort zone limits, except as follows. The operative temperature may go above or below the comfort zone limits during drift or ramp provided the drift or ramp starts inside the comfort zone at an operative temperature at least 0.6℃/h away from the limit that is exceeded, and the limits of the comfort zone are not exceeded for a period of more than 1.0 hour.

Non-Uniform Conditions

[15] Vertical Air Temperature Difference. In an enclosed space, air temperature generally increases from floor to ceiling. If this temperature difference is sufficiently large, a person's head can have local warm discomfort and/or cold discomfort at the feet even though the overall average is thermally neutral. To prevent this local discomfort, the standard calls for a maximum temperature difference between head and feet of 3℃.

[16] Radiant Temperature Asymmetry. **Being surrounded by surfaces that have large temperature differences is a frequent cause of discomfort, even when the air temperature is considered in the comfort zone.** These conditions are frequently caused by cold or hot windows, un-insulated walls or ceilings, direct sunlight, or improperly installed ceiling heating panels. **In general, people are more sensitive to asymmetric radiation caused by a warm ceiling than that caused by hot and cold vertical surfaces. So the standard recommends that the temperature difference in opposite directions (asymmetry) in the vertical plane shall be less than 5℃, and in the horizontal plane, less than 10℃.**

[17] Floor Temperature. A floor that is too warm or too cold will cause occupants to complain of the thermal discomfort of their feet for people wearing shoes. The temperature of the floor, rather than the material of the floor covering, is the most important factor for foot thermal comfort. The standard gives the allowable range of floor temperature as above 19℃ and below 29℃.

Draft

[18] Unwanted local cooling of the body caused by air movement is defined as draft. Factors affecting the draft sensation include the air speed and the air temperature. Sensitivity to draft is greatest where the skin is not covered by clothing, especially the head region and the lower leg region. Some persons are more sensitive to air motion than others. Higher air

speeds may be more acceptable if the person has individual control of the local air speed.

Humidity and Comfort

[19] When Willis Carrier developed the scientific basis for air-conditioning in 1902, he was not worried about human comfort, he was trying to solve a color-printing problem at a Brooklyn publishing plant. To prevent the paper from changing size, he needed to maintain a constant amount of water vapor in the air. However, for human comfort, much of the time as noted on the Psychrometric chart above, the amount of humidity is not critical as long as it stays below about 55% relative humidity (more exactly-below a dew point of 16.7℃).

Mean Radiant Temperature and comfort

[20] What has been described above would lead us to believe that controlling air temperature and humidity are the primary factors in achieving comfort. Most people equate comfort with air temperature and humidity level. In other words, they equate comfort with "air" conditioning. This can be a misconception. **A comfort variable that can be just as important as air temperature is the temperature of the surrounding surfaces that is referred to as the mean radiant temperature (MRT) the average of the surrounding surface temperatures.**

[21] The above diagram is simplified into two dimensions, but shows the concept. The mean radiant temperature depends on one's position in the space. When the person moves nearer the warm window, the body's view angle of the warm window becomes larger, thus a higher mrt. Our body's surface (skin) temperature is about 32.2~33.03℃ and radiates heat to cooler surfaces and receives heat from warmer surfaces. For maximum comfort, the body prefers to be surrounded with air and surface temperatures that average 23℃. **When the air temperature and the mrt are equal, our body loses as much of its excess heat by radiation to surrounding cooler surfaces as it loses to the surrounding air by convection.** Sitting near a large glass window whose temperature is 37.8℃ or more will raise the mrt significantly. To maintain the same comfort level when the surface temperatures rise, the surrounding air must be cooler.

Comfort Indices

[22] The Operative Temperature. A more useful term for defining comfort is the operative temperature (top). This is calculated by averaging the space's air temperature and its mean radiant temperature. Note that the horizontal axis of the Psychrometric chart above is operative temperature, so the comfort envelope accounts for three comfort variables: air temperature, mean radiant temperature and relative humidity. The fourth environmental variable, air motion, shifts the envelope laterally. As the air motion increases, the envelope moves to the right, and vice-versa.

[23] Passive comfort a result of surface temperatures. Consider that we are most comfortable during those periods of the year when the surfaces of our enclosure are naturally around 23℃. The average interior surface temperature results from the average of the daily

high and low outdoor temperature being about that same value, 23℃. This occurs for various durations during the spring and fall every year. The building is in the passive mode. We do not need to intervene with an active system to create comfort conditions; it occurs naturally. The temperature of the air in the space takes on the value of the average of the enclosing surfaces. The surfaces naturally "condition" the air.

[24] Passive discomfort as a result of surface temperatures. As the daily outdoor average temperatures rise, passive strategies, like night ventilation, become ineffective and heat is driven inward, raising the temperature of the building's mass. Some or all of the inside surfaces have heated up to above the upper limits of the comfort envelope (above 26℃) and the air in the space warms to the new higher average surface temperature. If no active cooling is supplied, interior conditions rise to well beyond the comfort envelope. The animation below shows how the comfort variables and building envelope change during a daily cycle when outdoor conditions rise to 38.3℃ with no conditioning.

[25] Active control of comfort by controlling indoor air temperature (Air Conditioning). **Compare the natural conditioning of the space described above to what happens when we need to go to the active mode with "air" conditioning.** To counteract the heat gain, we fill the space with cooler air. This cool air intercepts the incoming heat; it is warmed, and returns to the conditioner to be re-cooled. **So, we occupy a box filled with cool air, colder than it would be if it didn't have to remove heat from the interior surfaces.**

[26] Passive control of comfort by controlling indoor surface temperatures. With this form of conditioning, the temperature of one or more surfaces (or panels) is/are controlled. For either heating or cooling, these panels can only deal with sensible heat and not humidity control. A separate system would be necessary for ventilation and humidity control (if needed). With radiant panels, usually in the floor or the ceiling, the majority of the heat (over 50%) is transferred by radiant exchange. Both heating and cooling can be accomplished by low energy means (assisted passive) as in the Skytherm system developed by Harold Hay and demonstrated at various locations (Fig. 4).

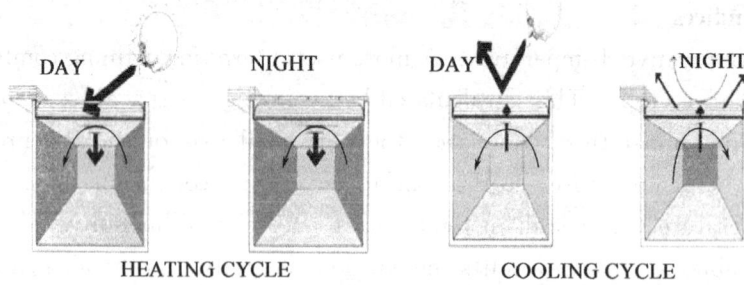

Fig. 4 Skytherm House

[27] In this 1979 house, solar heat is gathered in the roof's water bags during the day and conducts through the metal ceiling, providing a warm radiant panel in the spaces be-

low. Movable insulation panels above the water bags control the amount of heat retained. For summer cooling, the water bags are exposed to the cold night sky to reject heat and covered during the day. The cool ceiling absorbs heat from the occupied space.

[28] Active control of comfort by controlling indoor surface temperatures (Radiant Conditioning). For fully active radiant systems, we are most frequently seeing advances coming from Western Europe where higher energy cost spur more rapid development than in the United States. The ventilation and dehumidification systems are decoupled from the radiant system, providing economies of energy and superior control of both the sensible and latent loads. The heating and/or cooling loads of the radiant panels are served by pumps rather than less efficient fans used for conventional air systems. The smaller quantities of air require smaller fans and ductwork. The resulting system provides quieter operation and improved indoor air quality. Occupants seem to prefer the feeling of comfort obtained with a radiant system over that of the typical convective (or all-air) system. It may be more like that feeling of comfort we have in the spring and fall (21.7~23.9℃) when surfaces determine the air temperature.

[29] Design of building for passive comfort. The design of the passive systems in a building should attempt to maintain conditions in the comfort envelope as long as possible. The longer these conditions can be maintained passively during the year, the better the passive design. Extending this period is usually best accomplished with the use of thermal mass. During the passive period of operation in a thermal mass building, the value of the interior air temperature is primarily a result of the surrounding surface temperatures. So, a strategy of controlling surface temperatures (rather than air temperature) could be an alternative method of comfort control.

[30] The phrase, "warm floor, cold ceiling" or "warm feet, cool head" is very applicable to achieving comfort. The "warm floor/warm feet" applies particularly to the heating season, while the "cool ceiling/cool head" would be a summer strategy.

II. Words and Expressions

physiology	n.	生理学
indices	n.	指数,指标
Comfort Indices		舒适指数,舒适指标
thermal stress	n.	热应力
metabolic rate	n.	代谢率
sensible heat	n.	显热
latent heat	n.	潜热
thermoregulatory	adj.	体温调节的
homeostatic	adj.	自我平衡的
hypothalamus	n.	视丘下部
thermostat	n.	自动调温器,温度调节装置

dialated blood vessels　　　　　　n.　　　　主（动脉）血管
thermal sensations　　　　　　　　n.　　　　热感觉
clothing insulation value clothing　　　　　　衣服绝热值
met　　　　　　　　　　　　　　　　　　　代谢单位
Psychrometric Chart　　　　　　　　　　　湿度计算图，焓（温）湿图
sedentary　　　　　　　　　　　adj.　　　坐姿的
drifts　　　　　　　　　　　　　　n.　　　漂移
Ramps　　　　　　　　　　　　　n.　　　斜道，漂移
draft sensation　　　　　　　　　　　　　　吹风感
operative temperature　　　　　　　　　　　作用温度
view angle　　　　　　　　　　　　　　　　视角
radiant panels　　　　　　　　　　　　　　辐射板
ductwork　　　　　　　　　　　　n.　　　管道系统
passive comfort　　　　　　　　　　　　　被动舒适
mean radiant temperature　　　　　　　　　平均辐射温度
night ventilation　　　　　　　　　　　　　夜间通风

III. Notations

1. This section gives an overview of how human thermal comfort is affected by the body's physiology, and the external conditions surrounding the body.

　　how 引导的从句为介词 of 的宾语

本节概述了人体的生理现象和身体周围的外部环境条件如何影响人体的热舒适性。

2. That energy which is not converted to do work is in the form of heat, and if not needed to maintain a constant body temperature, it is brought to surfaces by blood flow (See Fig. 1), then rejected to the body's surrounding environment.

　　it 为 That energy

没有被转化为功的那部分能量以热的形式存在，如果没有被用于保持恒定的体温，则它被血液流动带到体表面（见图1），然后被释放到身体的周围环境中去了。

3. Over the years, researchers of human comfort have established the variables that affect a human's thermal sensationsand they have established the ranges of these variables within which the average person is comfortable.

　　within which 引导的从句修饰 ranges

多年来，从事人体舒适度研究的人员已经确定了一些影响人体热感觉的变量，并给出了其使一般人感觉舒适的取值范围。

4. One hundred years ago, Willis Carrier, developed a method that allows us to visualize two of the variables — the combination of air temperature and relative humidity that exist in a space.

　　the combination 为 variables 的同位语

一百年前，威利斯开利发展了一种方法，它能使我们直观地将存在于一个空间的诸变

量中的两个变量——空气温度和相对湿度的组合可视化地表现出来。

5. Being surrounded by surfaces that have large temperature differences is a frequent cause of discomfort, even when the air temperature is considered in the comfort zone.

被动分词短语　Being surrounded by...　作主语

被具有较大温差的表面包围是引起不舒适的一种常见原因，即使是空气温度处在舒适区。

6. In general, people are more sensitive to asymmetric radiation caused by a warm ceiling than that caused by hot and cold vertical surfaces. So the standard recommends that the temperature difference in opposite directions (asymmetry) in the vertical plane shall be less than 5℃, and in the horizontal plane, less than 10℃.

注意 more... than... 短语，第一个 that 为 asymmetric radiation

一般情况下，人们对由热天花板所致的不对称辐射比由冷热垂直表面引起的不对称辐射更为敏感。因此，标准建议，在垂直面相反方向（不对称）的温差应低于5℃，而在水平面，应低于10℃。

7. A comfort variable that can be just as important as air temperature is the temperature of the surrounding surfaces that is referred to as the **mean radiant temperature (MRT)**-the average of the surrounding surface temperatures.

第二个 that 为 temperature

一个与空气温度一样重要的舒适性变量是被称为平均辐射温度（MRT）的周围表面温度——周围表面温度平均值。

8. When the air temperature and the mrt are equal, our body loses as much of its excess heat by radiation to surrounding cooler surfaces as it loses to the surrounding air by convection.

注意 as... as... 短语的应用

当空气温度与平均辐射温度相等时，我们的身体通过辐射散失到周围冷表面的余热量与通过对流散失到周围空气的余热量一样多。

9. Compare the natural conditioning of the space described above to what happens when we need to go to the active mode with "air" conditioning.

compare... to 把…与…比较

将上述所述的空间自然调节情况与采用主动的"空气"调节模式的情况进行比较。

10. So, we occupy a box filled with cool air, colder than it would be if it didn't have to remove heat from the interior surfaces.

注意 it would be... if it didn't have to remove... 用法 如果…就会…

所以，用冷空气来充满整个空间，如果它没有被用于消除内表面的热量那么它将会变得越冷。

IV. Exercises

1. Translate the following sentences into Chinese.

(1) Thermal comfort is defined as that condition of mind that expresses satisfaction

with the thermal environment. Dissatisfaction may be caused by warm or cool discomfort of the body as a whole or may be caused by an unwanted cooling (or heating) of one particular part of the body.

(2) If it generates more heat than is needed, it must lose heat to its surroundings or its temperature will rise and it will become ill and could die. Likewise, if it loses too much heat to maintain a constant temperature, its temperature will lower and it could die.

(3) The human body can be compared to a machine (in the engineering world) that converts fuel into energy for the purpose of doing work — the more active the body, the more fuel that is consumed.

(4) Drifts or ramps are allowed without further restriction provided that the operative temperature does not go above or below the comfort zone limits, except as follows. The operative temperature may go above or below the comfort zone limits during drift or ramp provided the drift or ramp starts inside the comfort zone at an operative temperature at least 0.6℃/h away from the limit that is exceeded, and the limits of the comfort zone are not exceeded for a period of more than 1.0 hour.

(5) What has been described above would lead us to believe that controlling air temperature and humidity are the primary factors in achieving comfort. Most people equate comfort with air temperature and humidity level. In other words, they equate comfort with "air" conditioning.

(6) However, for human comfort, much of the time as noted on the psychrometric chart above, the amount of humidity is not critical as long as it stays below about 55% relative humidity (more exactly-below a dew point of 16.7℃).

(7) As the daily outdoor average temperatures rise, passive strategies, like night ventilation, become ineffective and heat is driven inward, raising the temperature of the building's mass.

(8) The phrase, "warm floor, cold ceiling" or "warm feet, cool head" is very applicable to achieving comfort. The "warm floor/warm feet" applies particularly to the heating season, while the "cool ceiling/cool head" would be a summer strategy.

2. Translate the following sentences into English.

(1) 要了解如何最好地控制热舒适，首先要了解人体是怎样保持热量平衡的。

(2) 描述热舒适的变量包括空气温度，相对湿度，空气流动，平均辐射温度。平均辐射温度是人体周围的所有围护结构内表面的平均温度。这四个变量被称为"环境变量"，因为它们代表了身体的周边环境。

(3) 一个更为有用的关于热舒适定义的术语为作用温度，它可以通过计算室内空气温度和平均辐射温度的平均值得到。

(4) 通风和除湿系统与辐射系统解耦，结果既可节能又可很好地对显热和潜热负荷进行控制。

(5) 利用辐射板，通常为地板或天花板，热量的大部分（超过50%）是通过辐射方式传递的。

Extensive Reading

Top 10 Ways Homeowners Can Ensure Good IAQ

[1] Vent bathrooms, kitchens, toilets and laundry rooms directly outdoors. Use energy efficient and quiet fans.

[2] Avoid locating furnaces, air conditioners and ductwork in garages or other spaces where they can inadvertently draw contaminants into the house. Install a door closer to ensure doors between houses and garages do not accidentally stay open. If ducts must pass through a garage or other potentially polluted space, seal them well to avoid entrainment of polluted air.

[3] Properly vent fireplaces, wood stoves and other hearth products; use tight doors and outdoor air intakes when possible.

[4] Vent clothes dryers and central vacuum cleaners directly outdoors.

[5] Store toxic or volatile compounds such as paints, solvents, cleaners and pesticides out of the occupiable space.

[6] Minimize or avoid unvented combustion sources such as candles, cigarettes, indoor barbecues, decorative combustion appliances or vent-free heaters.

[7] Provide operable windows or additional mechanical ventilation to every space to accommodate unusual sources or high-polluting events, such as the use of home cleaning products, hobby activities, etc.

[8] Use sealed-combustion, power-vented or condensing water heaters and furnaces. When natural-draft applications must be used, they should be tested for proper venting and should be located outside the occupiable space when possible.

[9] Put a good particle filter or air cleaner in your air-handling system to keep dirt out of the air and off of your ductwork and heating and cooling components. Maintain it or replace it regularly as required.

[10] Distribute a minimum level of outdoor air through-out the home, using whole-house mechanical ventilation.

A Brief History of Particle Image Velocimetry

[1] While the method of adding particles or objects to a fluid in order to observe its flow is likely to have been used from time to time through the ages no sustained application of the method is known. The first to use particles to study fluids in a more systematic manner was Ludwig Prandtl, who did so in the early 20th Century.

[2] Laser Doppler Velocimetry predates PIV as a laser-digital analysis system to become widespread for research and industrial use. Able to obtain all of a fluid's velocity measurements at a specific point, it can be considered the 2-dimensional PIV's immediate predecessor. PIV itself found its roots in Laser speckle velocimetry, a technique that sever-

al groups began experimenting with in the late 1970s. In the early 1980s it was found that it was advantageous to decrease the particle concentration down to levels where individual particles could be observed. At these particle densities it was further noticed that it was easier to study the flows if they were split into many small 'interrogation' areas, which could be analyzed individually to generate one velocity for each area. The images were usually recorded using analog cameras and needed immense amount of computing power to be analyzed.

[3] With the increasing power of computers and widespread use of CCD cameras it became tempting to do everything digitally. The implication of doing so was analyzed during 1990s and over time digital PIV became increasingly common, to the point that it today totally dominates.

Part 2　Ability Enhancement of Occupational English for HVAC
（专业英语应用能力拓展）

Lesson 16　English Writing Guideline（英文科研论文写作简介）

　　科技英语论文写作的目的是把作者的研究工作及成果呈现给全世界相同及相关领域的研究人员，一篇优秀的学术论文必须具有创新性、丰富的信息量、可读性强和优良的文字表达能力，对于英文论文写作，应注意学习规范的英文写作方法，以提高论文的可读性，更有效地体现其研究价值。

1. 科研论文的一般格式

　　科研论文一般具有较为固定的格式，英文科技论文的内容和格式可参见表1，不同期刊的格式和要求略有不同，具体要求请一定参考拟投期刊编辑部的来稿须知。

科研论文格式及其特点　　　　　　　　　　　　　　表1

组成部分名称（按文章顺序）	特点或简要说明
题目 Title	10～20 个字 10～20 words 简明，不必求全 Brief. A complete sentence is not necessary.
作者信息 姓名 单位地址 联系方式：E-mail 地址、传真、电话 Authorship Names of authors Affiliation E-mail address and telephone and fax numbers for corresponding author, if possible	通讯作者：往往是固定研究人员或项目负责人。 Corresponding author: Faculty member or principal investigator.
摘要 Abstract	150～200 英文词，说明研究目的、方法、结果、结论和意义。可以写一些定量结果。不仅对读者，而且对文献检索者都有帮助。 150～200 words to give purpose, methods or procedures, new results and their significance, and conclusions. Write for literature searchers as well as Journal readers. Include major quantitative data if they can be stated briefly, but do not include background material.
关键词 Key words	3～5 个关键词，作为论文检索用，使读者可用关键词方便地检索到此论文，并对论文按内容分类。 3～5 key words which can be regarded as index words.
符号表 Nomenclature, Notation or Symbols	说明文章中符号表示的量的意义，单位，尽量用国际单位制。 Use SI units as much as possible.
引言 Introduction	篇幅：全文的 10%～20%。 说明所研究问题的重要性；相关研究回顾与综述；指出已有研究的不足和局限，但语气应友善而含蓄。说明本论文的目的和重要性。 Introduce the importance of the problem studied. Review of previous work.

continued

组成部分名称（按文章顺序）	特点或简要说明
引言 Introduction	State the limitations or short comings of the previous work. Clearly state the purpose and significance of the present work. 注意：不必提及所有文献。如果一篇文献对所讨论专题做了综述，可只引用该文献，不必再重复标注所讨论的文献。一般情况下，引言不应超过3页（间行打印，无图表）。 Notice： Do not attempt to survey the literature completely. If a recent article has a survey on the subject, cite that article without repeating its individual citations. In general, the Introduction should be no more than 3 double-spaced word-processed pages with no figures and tables.
研究或试验方法 Research approach Theoretical section or Experimental section	篇幅：全文的 20%～30%。 介绍为简化问题所作的必要且合理的假设； 对问题进行数学描述：列方程、边界条件和初始条件； 问题求解； 或介绍试验仪器、条件和步骤，使读者阅读后可重复试验。 Make necessary assumptions. Describe the problem in mathematical equations together with relating boundary and initial conditions. Obtain the solution. Let the research can be reproduced. -Describe the apparatus and instruments. -Describe the pertinent and critical factors involved in the experimental work.
结果和讨论 Results and discussion	篇幅：全文的 40%左右。 研究结果介绍，数据的必要解释，新发现的讨论，与其他相关结果的比较。 结果和讨论也可分开。 结果：直接的发现；讨论：间接的发现。 此部分内容安排要特别注意逻辑性。 Present the results. Discuss new findings. Provide explanations for data. Elucidate models. Compare the results with other related words. Results and Disscussion may be separated. Results：direct findings. Discussion：indirect findings. Notice：please logically arrange the contents.
结论 Conclusions	介绍研究工作的主要结论，力求简明。 Summarize conclusions of the work.
致谢 Acknowledgement（s）	说明本工作受到的资助及得到的帮助。 Information regarding the supporter（s）(e. g., financial support) is included here.
参考文献 References	对于一般科研论文，参考文献为 10～20 篇；对于综述性论文，参考文献为 60～100 篇。 10～20 references for research paper and 60～100 references for review paper.
附录 Appendix	一些公式的详细推导等内容可放在附录部分，以便使论文更紧凑。 Some detailed derivation of equations etc. could be placed in this part.

2. 英文摘要的写作

（1）摘要及其分类

摘要是文献全文的浓缩，它提供了该文献的信息内容，摘要本身给读者一个信息，即该篇文献所包含的主要概念和讨论的主要问题，帮助科技人员决定这篇文献对自己的工作是否有用，也是国际知名检索机构收录该文献的重要依据。按美国工程信息公司编辑部（EI 编辑部）的分类，文摘分为指示性文摘与信息性文摘，或者两者结合。

指示性摘要（Indicated Abstract）仅指出文献的综合内容，适用于综述性文献、图书介绍及编辑加工过的专著等。

信息性摘要（Informative Abstract）多用于科技杂志或科技期刊的文章，也用于会议论文及各种专题技术报告。信息性摘要包括了原始文献的某些重要梗概，一般由主要研究目的、研究过程及方法和研究结果三部分组成：研究目的主要说明作者写此文章的目的，或说明本文主要要解决的问题；研究过程及方法则主要说明作者主要工作过程及所用的方法，也包括边界条件、使用的主要设备和仪器；研究结果是述及作者在此工作过程最后得出的结果和结论，如有可能，尽量提一句作者结论和结果的应用范围和应用情况。

（2）摘要的写作技巧

为了便于读者在没有机会阅读全文时了解该研究的目标、过程及成果；还便于文摘、索引等第二次出版物的转载，摘要必须具有语言概括性、内容完整性、主题突出性等特点。因此，摘要的写作着重注意文摘题目、主要概念词、文体风格、摘要长度等写作问题。

1）文摘的题目

对于文摘的题目，应力求简单明了直接反应文献的主题（一般不应超过 20 字），同时应注意以下几个方面的问题：

① 英文题目开头第一字不得用 The 、And 、An 和 A；

② 英文题目第一个字母大写，其余小写，下列情况除外：专用名词首字母大写；德语名词第一个字母应大写；

③ 文献的主副标题（题目）必须用句号分开，不得用分号或破折号；

④ 题目中尽量少用缩略词，必用时亦需在括号中注明全称（尽管中文文献题目中常用英文缩略字或汉语拼音首字母缩略字）；

⑤ 特殊字符即数学符号和希腊字母在题目中尽量不用或少用。

2）文体风格

摘要中要注意文体的风格（styles），可努力做到以下几个方面的要求：

① 文摘叙述要简明，逻辑性强；

② 句子结构严谨完整，尽量不用长句、复合句，而应使用简明、直接的短句子；

③ 技术术语尽量用工程领域的通用标准；

④ 用过去时态叙述作者工作，用现在时态叙述作者结论，如"The EER under different load for cooling mode was calculated, and the result shows that EER under the range of load ratio from 50% to 80% is higher than that under other load ratios."；

⑤ 可用动词的情况尽量避免用动词的名词形式，如"Thickness of plastic sheets was measured."而不用"Measurement of thickness of plastic sheet was made"；

⑥ 注意冠词用法，分清"a"是泛指，"the"是专指，如"Pressure is a function of temperature"而不应是"Pressure is a function of the temperature."；

⑦ 不使用俚语外语表达概念，应该用标准英语；

⑧ 尽量应用主动语态代替被动语态，如"A exceeds B"比"B is exceeded by A"好；

⑨ 组织好句子，使动词尽量靠近主语，例如不用"In this report, an efficient algorithm that employs the concept of static equilibrium to determine the stability of clamping is developed."而使用"In this report, we develop an efficient algorithm is developed that employs the concept of static equilibrium to determine the stability of clamping."；

⑩ 用重要的事实开头，尽力避免用辅助从句开头，例如"Power consumption of the VRF air-conditioning system was determined from data obtained experimentally."而不用"From data obtained experimentally, power consumption of the VRF air-conditioning system was determined."；

⑪ 删繁从简，如用"increased"代替"has been found to increase"，尽量不要使用"not only... but also"用"and"就行了；

⑫ 文摘中涉及其他人的工作或研究成果时，尽量列出他们的名字；

⑬ 文摘词语的拼写用英美拼法都可，但在每一篇中须保持一致；

⑭ 具有独立性，不能是引言和结论的简单重复。

(3) 摘要长度

一般应不超过 200 个词，不少于 100 个词。缩短摘要可采用以下常用的方法。

1) 取消不必要的字句：如删去"It is reported..."、"Extensive investigations show that..."、"The author discusses..."、"This paper concerned with..."、"... specially designed or formulated"等词语；

2) 对物理量单位及一些通用词可以适当进行简化：如不用 at a temperature of 250℃ to 300℃，而用 at 250~300℃；不用 at a high pressure of 2000 Pa，而用 at 2000 Pa；

3) 取消或减少背景情况（Background Information）；

4) 限制摘要只表示新情况、新内容，过去的研究细节可以取消；

5) 尽量简化一些措辞和重复的单元：如"本文所谈的有关研究工作是对过去老工艺的一个极大的改进"等切不可进入摘要；作者在文献中谈及的未来计划不纳入文摘。

3. 引言的写法

引言（Introduction）是对全文内容和结构的总体描述，主要任务是在读者阅读之前向读者介绍文章的主要内容和结构形式，以使读者快速、准确地阅读到所需内容。引言一般包含四方面的内容：研究背景，研究现状，提出问题和研究目的。下面以下列 3 个示例为例介绍引言的写法。

Study of free convection frost formation on a vertical plate

Frost formation processes are of great importance in numerous industricl applications including refrigeration, and process industries. In most cases, frost formation

背景资料	is undisable because it contributes to the increase in heat transfer resistance and pressure drop. Frost formation is a complicated transient phenomenon process in which a variety of heat and mass transfer mechanisms are simultaneous.
	Typical frost formation periods have been described by Hayashi et al. [1]: an initial one-dimensional (1-D) crystal growth is followed by a frost layer growth periods characterizes long time processes, in which the frost surface can reach the melting temperature. Each growth mode is characterized by peculiar values of frost density which in turn affects the other frost parameters (thickness, apparent thermal conductivity). Furthermore, as
文献回顾	observed in several studies [2-5], the features of the heat transfer rate (through the wall-to-air temperature difference) and of the mass transfer rate (which depends on air moisture content too) affect the frost structure and control the length of the growth periods. Owing to the complexity of the phenomenon, the development of reliable frost formation models as well as of correlations to evaluate frost properties is a demanding task; experimental data are required to check both the assumptions made in the theoretical analyses and the predicted results.
	As clearly reported in some review papers [6-8] frost formation during the forced convection of humid air has been extensively studied, while, on the other hand, only a limited number of investigations deal with mass-heat transfer during natural convection on a surface at subfreezing temperature. This problem was tackled by Kennedy and Goodman ([9], study of frost formation on a vertical surface), Tajima et al. ([10], flat surface with different orientations), Cremers and Mehra ([3], outer side of vertical cylinders), Tokura et al. ([4].
研究目的与活动	Vertical surface). To the author's knowledge, no data are available for the natural convection in channels, despite the practical significance of this phenomenon in such devices as evaporative heat exchangers for cryogenic liquid gasification.
	The present paper reports the results of an experimental investigation of frost formation on a vertical plate inside a rectangular channel where ambient air is flowing due to natural convection. The experiments have been conducted in the range of low-intermediate values of the relative humidity (31%～58%) for which frost temperature, during frost growth, is always below the triplepoint temperature, thus acting as a further parameter of the study. The measured data have been compared with the results of a mathematical model developed for predicting frost growth and heat flux at the cold plate.

(1) 研究背景的写法

研究背景（Study of free convection frost formation on a vertical plate）通常可以通过对某个现象或问题的描述，引出文章的研究目的，对于研究背景的写作，首先要介绍背景资料，一般包括 3 个步骤：

1）指出和论文内容所属的研究领域有关的一般事实；

2）转到大研究领域中的一个次研究领域，并指出该次领域中某些特定的事实；

3）焦点转到与该次领域中的论文所探讨的问题有密切关系的更狭窄的主题，并针对此主题再指出某些事实。在介绍背景时通常先有一两个范围比较广的句子，然后句子内容的范围逐渐缩小，最后一两句的范围应该相当狭窄。

介绍背景时需要注意时态，通常叙述有关某些现象或某个研究领域的普遍事实时主要动词用一般现在时态。如：① A few analytical models and simulations for conventional cylindrical heat pipe mainly concentrate on the startup operation involving liquid metal working liquids；② Heterogeneous（不同种类的）distillation is involved in several industrial process。而现在完成时态主要指出专业领域里最近发生的某种趋势，或最近发生的某个事件，如：Various approaches as modular technique and object oriented（导向的）programming have been employed as programming tools。

(2) 研究现状的写法

作为引言的一个重要组成部分，研究现状通常要介绍已有研究成果，引用时主要有三种写法：

1）依照参考文献的主题与作者研究论文关系的密切程度排列（不密切——→密切）；

2）依照时间顺序（早——→晚）；

3）依照所要讨论的参考资料的不同类别（例如根据不同的研究路线）。

研究现状常用的句型有资料导向引述、研究程度描述、多作者导向引述和论文作者导向引述。

1）资料导向引述

资料导向引述一般放在研究现状中的首句，常常叙述与研究论文内容关联的一般事实，作者需把自己讨论的主题作为句子的主语。常用时态是一般现在时和现在完成时，通常有以下两种格式：

• In some climates, control of house dust mite allergen（过敏原）loads has been found to be possible by seasonal regulation of indoor relative humidity [2]。（参考文献的序号放在括号内）

• This hypothesis is supported by cell morphology（形态学）results in studies of myalgic trapezius muscles (Larsson et al., 1998; Kadi et al., 1998)。（参考文献作者的姓氏及出版年代置于句末括号内）

2）研究程度描述

研究程度描述即描述从过去到现在的一种研究趋势，常用现在完成时。讨论的焦点是研究主题，而不是从事研究工作的学者。所以常用被动语态。There has been... there have been... 为常见句型，指出对某个问题的研究程度。如：

• There has been much research on the documentary structure of source code.
• There have been many studies on the documentary structure of source code.
• There has been little research on load margin and critical bus determination.
• There have been few reports on load margin and critical bus determination.

下列表2～表4中列出研究主题位于不同位置时研究程度描述的标准句型，以供科研论文写作时参考之用。

研究程度描述的标准句型——研究主题在动词之后　　　表2

研究程度	动词（现在完成时，被动语态）	主　题
Much Little No A volume of	work research	has been carried out on has been done on has been performed on has been conducted on has been devoted on	Fuzzy（失真）voltage stability analysis
Much Little	attention	has been devoted to has been directed toward has been focused on	the heterogeneous distillation
Many A number of Several of Quite few	studies experiments	have been conducted on have been done regarding have been performed on have been published on	the effect of delamination in flat laminates（碾压）

研究程度描述的标准句型——研究主题在动词之前　　　表3

主　题	动词（现在完成时，被动语态）	研究程度	时间（可有可无）
The documentary structure of source code	has been	the subject of much research the subject of few studies	
The effects of density fluctuations on the changes of azimuth（方位）	have been	the focus of a great deal of research	in recent years in the last decade since 1995
The documentary structure of source code	has drawn has attracted	much attention much interest little attention	
The effects of density fluctuations on the changes of azimuth	have drawn have attracted		

研究程度描述的标准句型——主题在动词之前（程度表示与表3不同）　　　表4

主　题	动词（现在完成时，被动语态）及研究程度	时间（可有可无）
The documentary structure of source code	has been widely discussed has been extensively examined has been thoroughly investigated has seldom been discussed	since 1992 in the last decade in recent years
The effects of density fluctuations on the changes of azimuth	have been widely discussed have been extensively examined have been thoroughly investigated have seldom been discussed	

3）多作者导向引述

多作者导向引述指在研究现状开始或研究现状新段落开始前，作者用一句话讨论两个或两个以上作者曾经做过的研究工作，其常用句型如表5～表7所示。

多作者导向引述常用句型（主动语态）　　　　　　　　　　表 5

多　作　者	动词（现在完成时，主动语态）	主　题	参考引述
Many investigators Several researchers A number of authors Few writers	have studied have investigated have examined have explored have reported on have discussed have considered	the fast decoupled（减弱震波） continuation power flows（功率通量）	(3, 4, 5, 7)

多作者导向引述常用句型（被动语态）　　　　　　　　　　表 6

主　题	动词（现在完成时，被动语态）	多作者	参考引述
The effects of devices such as VDTs（视频显示终端）	have been studied have been investigated have been examined have been explored have been reported on have been discussed have been considered	by several authors by a number of authors by many investigators by few writers	(Brawn, 1982; Forman, 1986; McGee, 1988)

多作者导向引述常用句型（包含以 that 开头的从句）　　表 7

多　作　者	动词（现在完成时，主动语态）	That	研究结果	参考引述
Several researchers	have found have shown have reported on have suggested	that	Muscle（肌肉）aches and joint（关节）pain can be reduced by the use of adjustable workstation furniture	(Kleeman, 1988; Roberts, 1990; Paul, 1993)

4）作者导向引述

论文作者在资料导向引述、研究程度描述或多作者引述之后，通常需要用 2~3 个句子把焦点转到与自己研究相关的一些参考资料，以介绍并评论这些参考文献提出的研究结果，引述这些参考文献的句子称为"作者导向引述"，如表 8 和表 9 所示。

作者导向引述常用句型之一　　　　　　　　　　　　　　　表 8

作者姓及文献年代	动词（一般过去时，主动语态）	That	研究结果
Hedge (1982)	showed found noted reported suggested observed pointed out	that	The open office caused too many disturbances（干扰）and distractions（分心）

作者导向引述常用句型之二   表9

作者姓及文献年代	动词（一般过去时，主动语态）	主题	第二句为作者导向引述
Lee (1999)	studied examined investigated explored considered	The effect of environmental conditions on frost formation	He found that…

第一句只描述某个学者的研究活动，第二句才叙述研究结果。

（3）提出问题及研究目的的写法

提出问题通常只需要一两个句子指出与本科研论文研究目的相关的内容即可，可以为下列内容之一：

1）以前的学者处理得不够完善或尚未研究的重要题目；

2）过去研究所派生的并值得探讨的新问题；

3）以前的学者曾经提出两个或以上的互不相容的观点或理论，为了解决这些互为差异的观点或理论之间的冲突，必须开展进一步的研究；

4）过去的研究自然可以扩展到新的领域或题目，或者以前曾提出的方法或技术可得到改善或延伸到新的应用范围。

表10和表11中列出了指出问题时常用的基本句型，包括单句和复合句，以供科研论文写作时参考之用。

指出问题的基本句型之一——单句   表10

转折副词	指出过去研究的不足	研究主题
However	few studies have been done on few studies have been reported on few studies have been published on few researchers have studied no studies have investigated little research has been devoted to little attention has been paid to little information has been published concerning no work has been done on little literature is available on there is little literature is available on little is known about insufficient data are available on	The supercritical fluid extraction of oil from herbaceous（草本的）substrates

指出问题的基本句型之二——复合句   表11

连接词	主题一的研究程度	主题二的研究程度
Although, while	much research has been devoted to A much work has been done on A many studies have been published concerning A many researchers have investigated A much literature is available on A	little research has been done on B little attention has been paid to B little work has been published on B little information is available on B few researchers have studied B few studies have investigated B

指出问题也可用"but"或"yet"的并列复合句,如:① "Many researchers have investigated neural network and fuzzy logic based techniques, but little work has been published on the RL based techniques." ② "Much research has been done on the effect of vortices (漩涡) on the frost growth rate, yet few experimental studies have been conducted in the range of low humidity conditions."。

作者在介绍自己的研究目的与研究活动时,可以采用以下两种导向方式:论文导向,即当研究论文的主要目的是分析某个问题、提出某种论证或介绍新的方法或技术时,把论文本身当作强调的内容;研究导向,即若研究论文主要提出某些实验或调查结果,则通常把研究活动作为步骤四的重点,即采用所谓的研究导向方式,表12~表14中列出研究目的写作的几种方式的基本句型。

研究目的——论文导向基本句型之一　　　　　　　　　　　　　表12

论文导向(一般现在时)	研究目的或主题
The purpose of this paper is The aim of this report is The objective of the present paper is	To obtain quantitative data about the particles size distribution To model the dynamic behavior in the super-or-near critical CO_2 extraction
The present paper reports This report presents This thesis describes This paper discusses	Heat transfer enhancement of the herringbone (人字形) type micro-fin tube The complexity of some decision and function computation problems involving counterfactual (反事实的) formulas
This paper proposes This thesis describes This paper presents	A fossil-fuelled power plant simulator with an ITS An intelligent tutoring systems (ITS) to enhance the capabilities of a power plant simulator

研究目的——论文导向基本句型之二　　　　　　　　　　　　　表13

论文导向(一般将来时)	研究主题
This paper will propose This thesis will present This paper will evaluate This paper will discuss	A new method for analyzing A Several approaches to improving A A theory that attempts to explain A New equations for expressing A
This paper will argue This report will present evidence to show This letter will present a proof In this paper, we will argue In this report, we will attempt to show	That Chen's assumption is false That the conventional method causes errors in special cases How these material variables affect paste formation during mixing

研究目的——研究导向的常用句型　　　　　　　　　　　　　表14

研究导向(一般过去时)	研究主题
The purpose of the experiment-reported here was The aim of this study was The objective of this research was	To determine the preferred viewing distance for work at a visual display unit To examine the apparent osmotic (渗透) behavior of some contemporary (当代的) commercial reverse osmosis membranes

研究导向	研究活动（一般过去时）
In this research described here In this study In the present investigation In this research In the experiments reported here	We determined a new method for reducing variance in common case Vertical nutrient mixing in late summer in test area 20 kilometers off the coast was examined

4. 研究或试验方法

（1）理论分析的写法

理论分析是科技论文所必需的组成部分，理论分析部分一般应包括数学分析、模型介绍和时态与语态。

模型介绍包括以下几部分：

1）模型的背景或理论基础；

2）作者所用方法或所述类型的基本假设；

3）基本方程式与分析；

4）模型的详细描述：说明或讨论模型的重要特点；

5）说明在特定情况下如何应用此模型或讨论一实际例子；

6）对例子中的解答进行评论。

理论分析部分常用一般现在时，往往采用第一人称复数开头的主动语态。如："It is supposed that the streamline passing from the center of the inflow to the center of the outflow corresponds to $\theta=0$."会显得比较生硬，而"We supposed that the streamline passing from the center of the inflow to the center of the outflow corresponds to $\theta=0$."就比较自然。

理论分析常常用到如下几种句型形式：

1) Consider the case in which the dynamic link parameters are unknown；

2) Given that $m=1.2$, we obtain... （列出方程式）；

3) The equation for at can be expressed as... （列出方程式）；

4) The relationship between m and n is as follows:... （列出方程式）；

5) If the material is used, the above equation reduce to （归纳为）.... （列出方程式）。

（2）试验方法的写法

试验不仅能让同领域的研究者在必要时也能重复和验证作者的研究方法，而且证明作者所采用的方法是经过认真仔细考虑的，能被认可的正确方法。因为本领域的研究者会非常仔细地审读试验与方法部分，以对作者的研究工作质量（可信性、有效性等）作出判断。

试验的基本内容是详细介绍所采用的材料、仪器仪表、设备及测试系统且清楚说明试验程序。如有必要，基本内容还可以包含以下资料：1）对整个试验的概述；2）选用特定材料、设备或方法的理由；3）试验的特殊条件或工况，如特殊的温度、大电流、紫外线辐射等；4）特殊试验设备或方法的详细介绍。标准方法或设备，则做简略介绍；5）应用的统计、分析方法的描述。

在叙述试验时往往要注意语态的使用，在叙述试验方法或步骤时往往使用被动语态，当作者想使读者特别注意自己的角色，例如强调自己的假设或建议、解释某种目的时，会把"we"当作句子的主语。

5. 结果和讨论

（1）结果的写法

结果部分是介绍自己的研究成果，并对研究成果进行评论或做出总结。介绍研究结果时往往以一个句子支出反映完整内容的图标，概述最重要的研究结果，并对研究结果做出评论或概括由这些结果得到的推论。

作者在介绍研究成果时常常必须叙述的内容包括：1）某个参数或变量在某段时间内的变化情况；2）不同试样、方法或研究对象之间的比较；3）不同参数或变量之间的关系或影响。

表 15～表 18 列出一些常用的基本句型：

表达参数或变量在某段时间内变化的基本句型 表 15

参数或变量	一般过去时动词	时间
The pressure	rose fell dropped	After more heat flux was added
The number of postgraduates in management school	declined remain constant remain unchanged	From 1998 to 2000

不同试样方法或研究对象之间比较的基本句型 表 16

项目1	用与比较的动词短语	项目2
The power throughput of the condenser	Increased much slower than	that of evaporator
The fastest algorithm（运算法则）was the genetic（遗传）algorithm		
Sample 1 had the highest magnetic resistance		

此部分常出现形容词、副词的比较级和最高级。

表达不同参数或变量之间的关系或影响的基本句型（简单句） 表 17

参数或变量1	关系	参数或变量2
Pressure	was correlated with was negatively correlated with was dependent on was independent of was determined by was closely related to	The ambient temperature

表达不同参数或变量之间的关系或影响的基本句型（复合句） 表 18

参数或变量 1	主句动词	连接词	参数或变量 2	从句动词
Pressure	increased decreased	as when	temperature	increased decreased

作者在叙述研究结果时，往往对研究结果进行说明或评论，包含的内容如下：1）根据本人的研究结果作出推论；2）作者解释研究结果或说明产生研究结果的原因；3）作者对此次研究结果与其他研究者曾发生的结果作比较；4）作者对自己的研究方法或技术的性能与其他研究者的方法、技术的性能进行比较；5）作者指出自己的理论模型是否与试验数据符合。

该部分也要注意时态的使用，在叙述自己的成果时，常用一般过去时，而描述由数学模型及理论分析得出的结果时，往往用一般现在时。当评论的内容为据研究结果作出的推论时，句子的主要动词常使用一般现在时，且常用 appear, suggest, seem 等推测动词或情态动词，如：These findings suggest that…；The contact angles may have effect on…；It appears that…。

当评论的内容为对研究结果可能的证明时，句子的主要动词之前通常加上 may 或 can 这些一般现在时态的情态动词。如：1）The layer structure or some other mixed complex material may be the most suitable refrigerants. 2）The results can be explained by… 3）One reason of this advantage may be that… 4）This may have occurred because…。

作者认为对结果的说明具有普遍有效性，则从句中使用一般现在时态；若作者比较不确定此说明是否具有普遍有效性时，则使用 may 加动词原形的形式；当作者认为说明范围只限在自己研究的特定情况下，则从句中使用一般过去时态。举例如下：1）It seems that ash deposition accelerates mainly through thermophoresis（热迁移）；2）It seems that ash deposition may accelerates mainly through thermophoresis；3）It seems that ash deposition accelerated mainly through thermophoresis。以上各句的有效性范围逐句缩小。

当评论的内容为比较作者的研究结果与其他学者的研究结果时，应该使用一般现在时态，因为不管两组结果之间是不是一致，这个比较是一种不受时间影响的逻辑上的关系，如：These results agree well with the findings…；These data are consistent with earlier findings…。

当评论的内容是对作者自身采用的方法或技术的性能与其他研究者提出的方法、技术的性能进行比较时，可使用一般现在时态或一般过去时态。若在评论中使用一般现在时态，则意味着作者认定句子的内容具有普遍有效性。若使用一般现在时态，则意味着作者认定句子的内容仅在特定情况下才有效。当作者指出由自己的理论模型所得到的预测与试验数据之间的吻合程度时，通常用一般现在时态，说明模型预测与试验数据是不是一致不受时间影响，如：1）The data confirm closely to the prediction of the model；2）The data indicate that the model is reliable and accurate；3）The model fits the experimental data well.

作者在叙述试验方法或步骤的句子往往使用被动语态，当作者想使读者特别注意自己的角色，例如强调自己的假设或建议、解释某种目的时，会把"we"当作句子的主语。

（2）讨论的写法

讨论部分的目的是阐明作者研究结果的意义，在该部分需说明自己的研究结果和自己研究领域中其他学者的研究之间的关系。同时要注意，该部分是论文中很关键的部分，应该直接表达自己对研究结果的解释和说明，千万不要过于含蓄。为了使自己的研究论文写得成功，作者不仅需要得出一些有意义的研究结果，而且还需要清楚地阐明这些结果的重要性，并说明这些结果值得其他学者注意的原因。

讨论部分的内容和结构如下：

1) 研究目的——再次概述自己研究的主要目的或假设；

2) 对结果的说明——对自己的研究结果提出说明、解释、猜测；

3) 推论或结论——指出依据自己的研究结果所得出的结论或推论；

4) 研究方法或结果的限值——指出自己研究的限值以及这些限值对研究结果可能产生的影响；

5) 研究结果的实际应用或建议新的研究题目——指出自己研究结果的实际应用及其价值，对进一步的研究方向或新的课题提出建议。

6. 结论

该部分的基本内容和结构如下：

（1）概述主要的研究工作（可有可无）；

（2）陈述研究的主要结论包括简略地重复最重要的发现或结果，指出这些发现或结论的重要内涵，对发现或结果提出可能的说明；

（3）研究结果可能的应用前景以及进一步深入的研究方向（可有可无）。

该部分的写作中也要注意以下几个事项：

（1）以十分清楚、简洁的方式叙述自己研究中的主要成果或发现，不要提及前文未涉及的新事实，也不要重复文中关于动机与试验程序的详细描述；

（2）部分研究论文以"conclusion"或"conclusions"作为标题，也有期刊用"concluding remarks"或"conclusions and future work"作为标题，作者最好查阅拟投期刊的惯用表达方式；

（3）在书写可能的应用时，不要过于勉强，特别是要避免夸大对效益的描述；

（4）区分结论与摘要。结论部分的功能在于叙述研究工作最重要的结果、结果的内涵、结果的说明等结论，而不是概述论文的所有内容。摘要则是对整个论文内容的概述，包括研究工作的目的、方法、主要结果及结论等信息。千万不要在结论中直接重复论文正文中其他部分的句子。

Lesson 17 Information of International Organization, Journals and Conferences (著名国际组织、国际期刊、国际会议)

1. HVAC 相关国际组织

- **American Society of Heating Refrigerating and Air-conditioning Engineers**，简称 **ASHRAE**

美国供热、制冷与空调工程师学会（ASHRAE）致力于促进供热、通风、空调、制冷及相关人性因素的科学和技术的研究与发展，以满足公众及 ASHRAE 会员对该领域科学和技术不断进步的要求。学会历史上由两个协会的管理者共同创建，即美国供热、通风工程协会（ASHVE）（1954 年改名为美国供热、空调协会（ASHAE））和美国制冷工程协会（ASRE）。这两个协会在 1959 年合并为美国供热、制冷与空调工程师学会（ASHRAE）。

其网址：www.ashrae.org

- **Air & Waste Management Association**，简称 **A&WMA**

空气和废弃物管理协会（A&WMA）由加拿大和美国的烟检人员始创于 1907 年，是一个非营利性专业组织。其成员包括科学家、工程师、决策者、律师以及政府、企业和大学的相关顾问。协会的宗旨是通过一个中立的信息交流论坛来发展改进环境科学与技术，完善相应的知识和法规，在重大的环境决策上为政府和社会提供帮助。

其网址：www.awma.org

- **International Society of Indoor Air Quality and Climate**，简称 **ISIAQ**

国际室内空气品质和气候协会（ISIAQ）始创于 1992 年。在第 5 届室内空气品质和气候国际会议（多伦多，1990 年）后，由 109 位国际科学家和从业者所创。

国际室内空气品质和气候协会是一个国际性的、独立的、多学科的非营利性组织，宗旨是通过以下方面为创造健康、舒适和有利于提高生产率的室内环境提供支持。

（1）推进室内空气品质领域科学和技术发展，内容涉及室内环境设计、施工、系统运行和维护以及空气品质测试和对人健康的影响；

（2）在室内空气品质和气候方面，通过出版和促进发行，推动国际和相应学科间的科技交流；

（3）组织、主办和赞助室内空气品质和气候方面的各种学术会议；

（4）为室内空气品质和气候的发展制定、改编和修订法规、标准和政策；

（5）与政府和其他部门以及对室内环境感兴趣的社团进行合作。

其网址：www.isiaq.org

2. HVAC 相关国际会议

- **ASHRAE**

由 ASHRAE 主办，每年两次，分为冬季年会（每年 1 月份在美国举办）和夏季年会

(每年6月份在美国举办)。冬季年会展览，展出一些暖通空调领域的新产品和新技术，有厂家参加。会后，一些会议论文编入 ASHRAE Transaction 中。

详细信息可登录 ASHRAE 网站：www.ashrae.org.

- **A&WMA 年会**

由 A&WMA 主办，每年一次。

详细信息可登录 A&WMA 网站：www.awma.org.

- **Indoor Air**

从1987年开始，三年一届的国际室内空气会议汇集了广大室内空气品质和气候的研究人员，致力于理解和解决室内空气品质和气候的相关问题。最新一届会议（Indoor Air 2008）在哥本哈根召开。2010年国际室内空气品质和气候会议（Indoor Air 2010）在中国北京举行。

详细信息请登录：www.indoorair2010.org/（第十一届国际室内空气品质和气候会议）

- **Healthy Building**

健康建筑国际会议起始于1988年，目的是将健康建筑的研究成果应用到实际中去。第一届健康建筑会议在瑞典首都斯德哥尔摩举办。此后，华盛顿、布达佩斯、米兰、奥斯陆和埃斯波、新加坡先后举办过此国际会议。最近一次会议于2009年在美国纽约州雪城举行。

会议的目的是将室内空气品质研究的最新知识、最新建筑技术以及产品开发应用到经济安全的健康居室和工作环境中去。

2009年会议信息可登录网站：www.hb2009.org/

- **ROOMVENT**

从创立开始，ROOMVENT 系列会议一直为从事空气流动的研究人员提供了良好的国际交流机会。会议召集了来自大专院校、国际研究协会以及工业界科学家和工程师来介绍和探讨各自领域的最新成果。

ROOMVENT 会议开始于1987年，由 SCANVAC（丹麦、芬兰、冰岛、挪威和瑞典组成斯堪的纳维亚暖通清洁工程协会联盟）在斯德哥尔摩举办。第二届 ROOMVENT 会议于1990年在挪威奥斯陆举办，从那以后，ROOMVENT 会议每两年举办一次，先后在奥尔堡（丹麦，1992）、克拉科夫（波兰，1994）、横滨（日本，1996）、斯德哥尔摩（瑞典，1998）、里丁（英国，2000）、哥本哈根（丹麦，2002）、可因布拉（葡萄牙，2004）、釜山（韩国，2009）和特隆赫姆（挪威，2011）。

详细情况请登录：http://www.sintef.no/Projectweb/Roomvent-2011/

- **ISHVAC**

由清华大学建筑技术科学系发起主办、中国国家自然科学基金资助的 ISHVAC（国际暖通空调研讨会）始办于1991年。以后每四年举办一次。最近一次会议即2011年第七届国际暖通空调研讨会在上海举行，由同济大学主办。会议的目的是聚集 HVAC 相关领域的国内外专家交流思想、理念以及研究成果，这些专家包括大学和研究协会的研究人员、工程师以及工业、政府部门的有关人员，有25场次的分会场交流，涉及建筑能耗及能效、室内环境及相关研究。

详细情况可登录：http://www.ishvac2011.org/

3. HVAC 相关国际期刊

- **ATMOSPHERIC ENVIRONMENT**

"大气环境"出版的论文主题涵盖了人与生态系统大气环境相互作用的所有方面。包括科学、管理、经济和政治等多方面的相互关系。"大气环境"的主要目的是为自然因素和人为因素对大气的影响提供科学证明。研究领域包括但不局限于以下一些方面：空气污染的研究及应用，空气品质及其影响，污染物传播和运输、沉积作用，生物-大气的物质交换，地球大气化学，辐射和气候等。基于试验的新论点、从局部的到全球范围的大气理论和模型都在本刊范围之内。

详细情况请登录：www.elsevier.com/locate/atmosenv

- **INDOOR AND BUILT ENVIRONMENT**

"室内及人工环境"出版下列方面的研究论文：室内和建筑物内的环境品质，以及它们对健康、工作、效率和人体舒适性可能造成的影响。论题范围还包括城市基建、建筑设计以及用于试验研究的材料（包括动物模拟和人造环境效果实验）。

详细情况请登录：http://ibe.sagepub.com/

- **BUILDING AND ENVIRONMENT**

"建筑和环境"出版的文章内容包括建筑研究及其应用的研究成果，内容涉及以下方面：

- 建筑物的环境特性、组成和原材料；
- 室内气候设计和性能；
- 人体对室内外物理环境的反应；
- 环境设计方法和技术，包括计算机模拟；
- 建筑设计、规划、政策制定的应用实例研究；
- 建筑环境方面的维护和再利用；
- 本地传统建筑和住宅的环境参数（包括经济、社会和文化等相关方面）；
- 国际技术和本地传统的综合；
- 建筑研究和建筑科学的原理和策略；
- 建筑科学和技术的历史；
- 建筑设计专业的教育情况。

详细情况请登录：www.elsevier.com/locate/buildenv

- **ENERGY AND BUILDINGS**

"能源和建筑"是出版与建筑物能源利用相关文章的国际期刊，目的是介绍建筑节能及提高室内环境品质方面的创新研究成果和试验结果。

涉及的方面有：

- 现有建筑和未来建筑的能源需求和消耗；
- 热舒适性和室内空气品质；
- 自然通风和机械通风；
- 空调建筑内的气流组织；
- 太阳能及其他可再生能源在建筑中的应用；

- 大空间建筑的能量平衡（工厂、公共建筑等）；
- 住宅、公共建筑及工业建筑的暖通空调和制冷系统；
- 建筑热回收系统；
- 建筑物和区域供热；
- 人工环境的能量守恒；
- 节能建筑；
- 建筑物理；
- 建筑围护结构和能耗；
- 室内冷热和照明系统的评价和控制；
- 智能建筑；
- 建筑设计和机械、照明系统的关系；
- 建筑新材料及其对能耗的影响；
- 节能建筑的内外部设计。

基于模拟得出结果的论文同样受到欢迎，尤其是那些与试验或实际测量有明显关系的将更受欢迎。

详细情况请登录：www.elsevier.com/locate/enbuild

- **INDOOR AIR-INTERNATIONAL JOURNAL OF INDOOR AIR QUALITY AND CLMAIE**

该刊是非工业建筑室内环境方面独创研究成果报道的平台。出版的论文包括以下方面：室内空气品质对健康的影响、热舒适性、监测和建模、污染源的描述、通风及其他控制技术等。研究结果将为建筑设计师、业主及经营者提供参考信息，以给居住者提供一个健康舒适的环境，也给医学家对怎样处理与室内环境相关的病症提供信息。

详细信息请登录：www.blackwellpublishing.com

- **ENVIRONMENTAL SCIENCE &TECHNOLOGY**

"环境科学与技术"是广大环境学科专业人员权威的信息来源之一，致力于出版自然和人工环境方面的科研文章，尤其关注环境中的化学作用，包括自然和人为两个方面，也包括生理现象以及与认识、控制环境相关的数学和计算方法，同样收录描述改善、控制和预防污染的重大技术进展的文章。其涉及内容如下：自然和人工环境的描述、环境的运转过程、环境的评价方法、改善和控制技术、可持续工程和绿色化学、政策以及文章评论。

详细情况请登录：pubs.acs.org/journals/esthag

- **ENERGY JOURNAL**

"Energy Journal"是 IAEE 的官方季刊。它创办于 1980 年，目的是促进能源及相关方面知识的进步和传播。编者们一直致力于出版一系列融合理论、实际经验和政策的文章。每期季刊共 150 页，包含与能源相关的研究文章、短讯及综述文章。包括的范围和主题有：

- 能源和环境问题；
- 石油；
- 电力行情；
- 发展中国家和能源；

- 天然气；
- 汽油需求分析；
- OPEC 和石油市场；
- 可再生能源；
- 方针政策；
- 煤炭问题；
- 分布的形成；
- 经济模型；
- 运输燃料的选择；
- 能量的效率；
- 经济学的规章制订；
- 能源税收；
- 市场力研究；
- 传统能源的取代；
- 核能问题；
- 运输；
- 散发物的交换（SO_2，CO_2）；
- CO_2 散发量的降低。

详细情况请登录：www.iaee.org/en/publications/journal.aspx

- **ENERGY**

该期刊是多学科综合的焦点，集中于能源相关项目的发展、评估和管理，主要包括：节能系统的输入输出分析、能量维持的测量及其实现、能源系统管理评价、环境影响评价、强调经济性的方案选择等。

详细情况请登录：www.elsevier.com/locate/energy

- **HVAC&R RESEARCH**

"International Journal of Heating, Ventilating, Air-Conditioning and Refrigerating Research"是英国建筑设备工程师协会创办的期刊，其文章的评定基于它对 R&D 的价值以及是否将研究描述得足够详细以对其他研究人员的工作产生帮助的程度。该刊始于 1995 年，出版的文章包含全球性的主题，如：影响室内空气品质的散发源、模拟制冷剂和润滑油混合物的热力学方法以及能量维持策略等。

详细情况请登录：www.ashrae.org/template/EducationLinkLanding;/category

- **JOURNAL OF THE AIR & WASTE MANAGEMENT ASSOCIATION**

"空气和废弃物管理协会期刊"是全球最老的环境技术期刊之一，最早以"Air Repair"的名字发行于 1951 年。该刊旨在为空气污染控制和废物处理相关专业人员提供服务，提供及时而可靠的信息。

详细信息请登录：www.awma.org/

- **ASHRAE JOURNAL**

"ASHRAE JOURNAL"是美国供热、制冷与空调工程师学会出版的期刊，主要面向应用研究，发表较宽范围内 HVAC&R 技术性研究论文。它所包含的内容从描述基本特性到综述新兴技术，包括了所有 HVAC&R 应用的范围。考虑到其权威性、平衡性和实用性的编辑内容，它是一本美国供热、制冷与空调工程师学会会员都会收到的月刊。其读者包括顾问工程师、机械承包人和建筑设计公司雇佣的工程师、建筑师、操作工程师和负责 HVAC&R 服务的车间工程师以及由 OEMs 与其他一些设计研究和开发的机构雇佣的设计工程师。

详细情况请登录：www.ashrae.org/template/JournalLanding

Part 3　Appendix（附录）

Appendix A Basic Terminology for HVAC（专业术语荟萃）

- 采暖、供热、锅炉与锅炉房工艺工程

1. 采暖 heating; space heating
使室内获得热量并保持一定温度，以达到适宜的生活条件或工作条件的技术，也称供暖。

2. 集中采暖 central heating; concentrated heating
热源和散热设备分别设置，由热源通过管道向各个房间或各个建筑物供给热量的采暖方式。

3. 全面采暖 Whole heating
为使整个采暖房间保持一定温度要求而设置的采暖。

4. 局部采暖 local heating
为使室内局部区域或局部工作地点保持一定温度要求而设置的采暖。

5. 连续采暖 continuous heating
对于全天使用的建筑物，使其室内平均温度全天均能达到设计温度的采暖方式。

6. 间歇采暖 intermittent heating
对于非全天使用的建筑物，仅在其使用时间内使室内平均温度达到设计温度，而在非使用时间内可自然降温的采暖方式。

7. 值班采暖 standby heating
在非工作时间或中断使用的时间内，为使建筑物保持最低室温要求而设置的采暖。

8. 热水采暖 hot water heating
以热水作热媒的采暖。

9. 高温热水采暖 high temperature water heating; high-pressure hot water heating
以温度高于100℃的热水作热媒的采暖，也称高压高温水采暖。

10. 蒸汽采暖 steam heating
以蒸汽作热媒的采暖。

11. 高压蒸汽采暖 high-pressure steam heating
以工作压力高于70kPa的蒸汽作热媒的采暖。

12. 低压蒸汽采暖 low-pressure steam heating
以工作压力低于或等于70kPa但高于当地大气压力的蒸汽作热媒的采暖。

13. 真空采暖 vacuum heating
工作压力低于当地大气压力的蒸汽采暖。

14. 对流采暖 convection heating
利用对流换热或以对流换热为主的采暖方式。

15. 散热器采暖 radiator heating
借助于散热器向室内传热以达到室温要求的采暖方式。

16. 集中送风采暖 localized air supply for air-heating

在一定高度上，将热风从一处或几处以较大速度送出，使室内造成射流区和回流区的热风采暖。

17. 热风采暖　warm-air heating；hot air heating

利用热空气作媒质的对流采暖方式。

18. 辐射采暖　panel heating；radiant heating

以辐射传热为主的采暖方式。

19. 地板辐射采暖　floor panel heating

以热水或热风作热媒，加热元件镶嵌在地板中的低温辐射采暖。

20. 煤气红外线辐射采暖　gas-fired infrared heating

利用可燃气体在辐射器中通过一定方式的燃烧，主要以红外线的形式放散出辐射热的高温辐射采暖。

21. 太阳能采暖　solar heating

通过一定手段，将太阳辐射能转换成热能的采暖。

22. 热媒参数　heating medium parameter

表征热媒状态的物理量，如供水温度、回水温度和供气压力等。

23. 供水温度　supply water temperature

水系统入口处的水温。

24. 回水温度　return water temperature

水系统出口处的水温。

25. 供热　heat supply；heating

利用热媒将热能从热源输送至各热用户的技术。

26. 区域供热　district heating；district heat supply

以热水或蒸汽作热媒，由热源集中向一个城镇或较大区域供应热能的方式。

27. 耗热量　heat loss

围护结构在室内外温差作用下向外传递的热流量，分基本耗热量和附加耗热量两部分。

28. 热水采暖系统　hot water heating system

以热水作热媒的采暖系统，一般分为自然循环和机械循环热水采暖系统两种。

29. 蒸汽采暖系统　steam heating system

以蒸汽作为热媒的采暖系统。

30. 采暖设备　heating equipment；heating appliance

泛指用于采暖的各种设备，如锅炉、换热器、暖风机、散热器等。

31. 锅炉　boiler

利用热能将水加热或使其产生蒸汽的热源装置。

32. 热水锅炉　hot water boiler

用于制取热水的锅炉。

33. 蒸汽锅炉　steam boiler

用于制取蒸汽的锅炉。

34. 换热器　heat exchanger

温度不同的流体在其中进行热量交换的设备，也称热交换器。

35. 水-水式换热器　water-water type heat exchanger
加热用的热媒和被加热的介质均为水的换热器。

36. 汽-水式换热器　steam-water type heat exchanger
加热用的热媒为蒸汽，被加热的介质为水的换热器。

37. 表面式换热器　surface-type heat exchanger; indirect heat exchaner
被加热的水与热媒不直接接触，而通过金属表面进行热交换的换热器，如壳管式、套管式、板式和螺旋板式换热器等，也称间接式换热器。

38. 汽-水混合式换热器　steam-water mixed heat exchanger; direct-coatact heat exchanger
使蒸汽和水直接接触进行混合而实现热交换的换热器，如淋水式、喷管式换热器等。

39. 蒸汽喷射器　steam ejector
直接利用高压蒸汽作为热源和动力源的一种换热、加压装置。

40. 膨胀水箱　expansion tank
热水系统中对水体积的膨胀和收缩起调剂补偿等作用的水箱。

41. 凝结水箱　condensate tank
蒸汽系统中用于汇集和贮存凝结水的水箱。

42. 开式水箱　open tank
与大气直接连通的水箱。

43. 闭式水箱　closed tank
不与大气直接连通的水箱。

44. 补给水泵　make-up water pump
特指向锅炉、热网和采暖系统补水用的水泵。

45. 循环泵　circulating pump
特指使水在锅炉、热网或采暖系统中循环流动的水泵。

46. 加压泵　booster
增加水系统作用压力的水泵。

47. 凝结水泵　condens
用于输送蒸汽凝结水的水泵。

48. 手摇泵　hand pump
人力驱动的水泵。

49. 真空泵　vacuum pump
能使封闭系统或容器产生一定真空度的设备。

50. 散热器　radiator; heat emitter
以对流和辐射方式向采暖房间放散热量的设备。

51. 分汽缸　steam manifold; steam header
蒸汽系统中，用于向各个分支系统集中分配蒸汽的截面较大的配汽装置。

52. 分水器　header
水系统中，用于向各个分支系统集中分配水量的截面较大的配水装置。

53. 集水器　header
水系统中，用于汇集各个分支系统回水的截面较大的集水装置。

54. 集气罐　air collector
用以聚集和排除水系统中空气的装置。

55. 同程式系统　reversed return system
热媒沿管网各立管环路流程相同的系统。

56. 异程式系统　direct return system
热媒沿管网各立管环路流程不同的系统。

57. 干管　main pipe；main；trunk pipe
连接若干立管的具有分流或合流作用的主干管道。

58. 立管　riser
竖向布置的热水或蒸汽系统中与散热设备支管连接的垂直管道。

59. 支管　branch pipe；branch
与散热设备进出口连接的管段。

60. 水力计算　hydraulic calculation
为使系统中各管段的流量符合设计要求，所进行的管径选择、阻力计算及压力平衡等一系列运算过程。

61. 环路　circuit；loop
特指流体可在其中进行循环流动的闭合通路。

62. 最不利环路　index circuit
系统中流体阻力最大的环路。

63. 管段　pipe section
特指系统中流量不发生变化的管道段落。

64. 管段长度　length of pipe section
管段实际延续的长度。

65. 当量长度　equivalent length
在系统的水力计算中，将局部阻力折算成与之相当的同一管径的摩擦阻力所对应的管段长度。

66. 折算长度　effective length
管段长度与当量长度之和。

67. 摩擦阻力　friction loss；frictional resistance
当流体沿管道流动时，由于流体分子间及其与管壁间的摩擦而引起的阻力。

68. 比摩阻　specific frictional resistance
单位长度管道的摩擦阻力。

69. 摩擦系数　friction factor
流体分子间及其与管壁间摩擦而产生阻力的无量纲数，也称摩擦阻力系数。

70. 绝对粗糙度　absolute roughness
管道内表面不规则起伏中的峰谷平均高差。

71. 相对粗糙度　roughness factor

管道的绝对粗糙度与该管道直径的比值。

72. 局部阻力　local resistance

当流体流经设备及管道中的三通、弯头等附件时，在边界急剧改变的区域，由于涡流和速度的重新分布而产生的阻力。

73. 局部阻力系数　coefficient of local resistance

流体流经设备及管道附件所产生的局部阻力与相应动压的比值。其值为无量纲数。

74. 当量局部阻力系数　equtvalent coefficient of local resistance

在系统的水力计算中，将摩擦阻力折算成与之相当的局部阻力所对应的局部阻力系数。

75. 折算局部阻力系数　effective coefficient of local resistance

局部阻力系数与当量局部阻力系数之和。

76. 阻力平衡　hydraulic resistance balance

通过计算并采取相应措施，使系统各并联管路在设计流量下的阻力差额率控制在允许范围内。

77. 压力损失　pressure drop

流体在管道及设备中流动时，由于摩擦阻力和局部阻力而导致的压力降低。

78. 极限流速　limiting velocity

在系统水力计算中所容许采用的流体最大流速。

79. 经济流速　economic velocity

在系统水力计算中，根据建设投资与运行费用、钢材消耗与动力消耗等因素，经技术经济比较确定的流体流动速度。

80. 系统阻力　system resistance

系统最不利环路的摩擦阻力与局部阻力之和。

81. 水力失调　hydraulic disorder

系统中各并联管路的实际流量与设计流量的偏差超过允许范围。

82. 作用半径　operating range

在一定压力作用下，系统的有效服务范围。

83. 资用压力　available pressure

可供用于克服系统中流体流动阻力的压力。

84. 静压　static pressure

(1) 流体在静止时所产生的压力；

(2) 流体在流动时产生的垂直于流体运动方向的压力。

85. 疏水器　steam trap

能从蒸汽系统中排除凝结水同时又能阻止蒸汽通过的装置。

• 通风工程

86. 通风　ventilation

为改善生产和生活条件，采用自然或机械方法，对某一空间进行换气，以造成卫生、安全等适宜空气环境的技术。

87. 工业通风　industrial ventilation

对生产过程中的余热、余湿、粉尘和有害气体等进行控制和治理而进行的通风。

88. 自然通风　natural ventilation
在室内外空气温差、密度差和风压作用下实现室内换气的通风方式。

89. 机械通风系统　mechanical ventilating system
为实现通风换气而设置的由通风机和通风管道等组成的系统。

90. 机械送风系统　mechanical air supply system
将室外清洁空气或经过处理的空气送入室内的机械通风系统。

91. 机械排风系统　mechanical exhaust system
从局部地点或整个房间把含有余热、余湿或有害物质的污染空气排至室外的机械通风系统。

92. 联合通风　natural and mechanical combined ventilation
自然与机械相结合的通风方式。

93. 局部送风系统　local air supply system；local relief system
为实现局部送风而设置的通风系统。

94. 局部排风系统　local exhaust system
为实现局部送风而设置的通风系统。

95. 事故通风　emergency ventilation
用于排除或稀释生产房间内发生事故时突然散发的大量有害物质、有爆炸危险的气体或蒸气的通风方式。

96. 局部送风系统　local air supply system；local relief system
为实现局部送风而设置的通风系统。

97. 局部排风系统　local exhaust system
为实现局部送风而设置的通风系统。

98. 通风设备　ventilation equipment；ventilation facilitis
为达到通风目的所需的各种设备的统称，如通风机、除尘器、过滤器和空气加热器等。

99. 送风机　supply fan
用于送风的通风机。

100. 排风机　exhaust fan
用于排风的通风机。

101. 通风机室　fan room；fan house
用于配置、安装通风设备的专用房间。

102. 送风机室　supply fan room
用于配置、安装通风设备的专用房间。

103. 排风机室　exhaust fan room
用于配置、安装排风设备的专用房间。

104. 进风口　air intake
采集室外空气的孔口。

105. 百叶窗　louver；shutter

由倾斜板条组成的窗式风口。

106. 局部排风罩　exhaust hood; hood
局部排风系统中，设置在有害物质发生源处，就地捕集和控制有害物质的通风部件。

107. 外部吸气罩　capturing hood
设在污染源附近，依靠罩口的抽吸作用，在控制点处形成一定的风速，排除有害物质的局部排风罩。

108. 接受式排风罩　receiving hood
设在污染源附近，利用生产过程中污染气的自身运行接受和排除有害物质的局部排风罩，如高温热源上部的伞形罩、砂轮机的吸尘罩等。

109. 密闭罩　exhausted enclosure; enclosed hood
将有害物质源全部密闭在罩内的局部排风罩。

110. 局部密闭罩　partial enclosure
仅将工艺设备放散有害物质的部分加以局部密闭的排风罩。

111. 整体密闭罩　integral enclosure
将放散有害物质的设备大部分或全部密闭起来的排风罩。

112. 大容积密闭罩　large space enclosure; closed booth
在较大范围内将整个放散有害物质的设备或有关工艺过程全部密闭起来的排风罩。

113. 排风柜　laboratory hood; fume hood
一种三面围挡，一面敞开或装有操作拉门的柜式排风罩。

114. 伞形罩　canopy hood
装在污染源上面的伞状排风罩。

115. 侧吸罩　lateral hood; side hood
设置在污染源侧面的排风罩。

116. 余热　excess heat; excessive heat
在不进行通风的条件下，室内得热量大于失热量的状况。

117. 余湿　moisture excess
在不进行通风的条件下，室内散湿量大于从室内排出的湿量的状况。

118. 穿堂风　through flow; through-draught; cross-ventilation
在风压作用下，室外空气从建筑物一侧进入，贯穿内部，从另一侧流出的自然通风。

119. 热压　thermal pressure; thermal buoyancy; stack effect pressure
由于温差引起的室内外或管内外空气柱的重力差。

120. 风压　wind pressure
风流经建筑物时，在其周围形成的静压与稳定气流静压的差值。

121. 粉尘　dust
由自然力或机械力产生的，能够悬浮于空气中的固态微小颗粒。国际上将粒径小于 $75\mu m$ 的固体悬浮物定义为粉尘，在通风除尘技术中，一般将 $1\sim 200\mu m$ 乃至更大粒径的固体悬浮物均视为粉尘。

122. 烟（雾）　fume
由燃烧或熔融物质挥发的蒸气冷凝后形成的，其粒径范围一般为 $0.001\sim 1\mu m$ 的固体悬浮粒子。

123. 烟气　fumes

在化学工艺过程中生成的通常带有异味的气态物质。

124. 液滴　droplet

在静止条件下能沉降，在湍流条件下能悬浮于气体中的微小液体粒子。

125. 雾　mist

悬浮于气体中的微小液滴，如水雾、漆雾、硫酸雾等。

126. 过滤效率　filter efficiency

过滤器所捕集的粒子质量或数量与过滤前空气中含有的粒子质量或数量之比，用百分率表示。

127. 吸附剂　adsorbent

具有较大吸附能力的固体物质。

128. 吸附质　adsorbate

吸附剂所吸附的物质。

- **空调工程**

129. 空气调节　air conditioning

使房间或封闭空间的空气温度、湿度、洁净度和气流速度等参数，达到给定要求的技术。

130. 舒适性空气调节　comfort air conditioning

为满足人的舒适性需要而设置的空气调节。

131. 工艺性空气调节　industrial air conditioning；process air conditioning

为满足生产工艺过程对空气参数的要求而设置的空气调节。

132. 空气调节机房　air conditioning machine room；air handling unit room

安装和运行空气调节设备的专用房间。

133. 显热　sensible heat

在物质的吸热或放热过程中，能使其温度发生变化的热量。

134. 潜热　latent heat

在一定温度和压力下，物质发生相变过程中，所吸收或放出的热量。

135. 全热　total heat

显热和潜热之和。

136. 综合温度　sol-air temperature

在计算空气调节房间外围护结构得热量时，所采用一种假想室外空气温度，在该温度的作用下进入围护结构外表面的热量等于在室外空气温度和太阳辐射共同作用下进入该外表面的热量。

137. 逐时综合温度　hourly sol-air temperature

综合温度的逐时值。

138. 逐时冷负荷　hourly cooling load

冷负荷的逐时值。

139. 房间得热量　space heat gain

进入和散入房间的热流量。

140. 人体散热量　heat gain from occupant
人体散热所形成的房间得热量。

141. 设备散热量　heat gain from appliance and equipment
设备与器具散热所形成的房间得热量。

142. 照明散热量　heat gain from lighting
灯具散热所形成的房间得热量。

143. 蓄热　heat storage; thermal storage effect
由于围护结构与家具等物体具有一定的热容量，而使房间产生对于得热量的蓄积和释放现象。

144. 蓄热特性　heat storage capacity; thermal storage characteristic
房间固有的蓄热放热能力，这种能力决定了房间阻抗热干扰的性能及得热与负荷之间的数量转换关系。

145. 散湿量　moisture gain
由湿源散入房间的湿流量。

146. 空气调节系统　air conditioning system
以空气调节为目的而对空气进行处理、输送、分配，并控制其参数的所有设备、管道及附件、仪器仪表的总合。

147. 气流组织　air distribution; space air diffusion
对室内空气的流动形态和分布进行合理组织，以满足空气调节房间对空气温度、湿度、流速、洁净度以及舒适感等方面的要求。

- 制冷工程

148. 制冷　refrigeration
用人工方法从一物质或空间移出热量，以便为空气调节、冷藏和科学研究等提供冷源的技术。

149. 制冷量　refrigerating effect
单位时间内，由制冷机蒸发器中的制冷剂所移出的热量。

150. 空调工况制冷量　rating under air conditioning condition
在规定的空气调节工作状况下，制冷机的制冷量。

151. （制冷）性能系数　(refrigerating) coefficient of performance (COP)
在制定工况下，制冷机的制冷量与其净输入能量之比。

152. 冷凝器　condenser
制冷剂蒸气在其中进行冷凝的换热器。

153. 蒸发器　evaporator
液态制冷剂在其中进行吸热蒸发的换热器。

154. 冷却塔　cooling tower
使循环冷却水同空气相接触，以蒸发的方式达到冷却目的的一种换热设备。

155. 热力膨胀阀　thermostatic expansion valve
用以自动调节流入蒸发器的液态制冷剂流量，并使蒸发器出口的制冷剂蒸气过热度保持在规定限值内的节流设备。

156. 贮液器　liquid receiver；receiver
制冷系统中贮存备用液态制冷剂的容器。

157. 冷凝压力　condensing pressure
制冷剂蒸气冷凝时的压力。

158. 水冷式冷凝器　water-cooled condenser
以水为冷却介质的冷凝器。

159. 风冷式冷凝器　air-cooled condenser
以空气为冷却介质的冷凝器。

160. 冷凝温度　condening temperature
制冷剂蒸气在冷凝器中冷凝时，对应于冷凝压力下的饱和温度。

161. 蒸发压力　evaporating pressure
制冷剂液体在蒸发器内蒸发时的压力。

162. 蒸发温度　evaporating temperature
制冷剂液体在蒸发器内气化时，对应于蒸发压力的饱和温度。

163. 吸气压力　suction pressure
压缩机进口处吸气管内制冷剂气体的压力。

164. 吸气温度　suction temperature
压缩机进口处吸气管内制冷剂气体的温度。

165. 排气压力　discharge pressure
压缩机出口处排气管内制冷剂气体的压力。

166. 排气温度　discharge temperature
压缩机出口处排气管内制冷剂气体的温度。

167. 标准工况　standard condition
符合标准规定的制冷机运行条件。

168. 冷水　chilled water
制冷机制出的低温水或天然冷源水。

169. 冷却水　cooling water
制冷装置的冷却用水。

170. 焓　enthalpy
物质所具有的一种热力学性质。定义为该物质的体积、压力的乘积与内能的总和。

171. 熵　entropy
一个热力状态函数。对于经历由状态1到状态2的可逆过程的封闭体系来说，其变化值用积分给定。

172. 㶲　exergy
能量中可能转换为最大有用功的部分。

173. 炕　anergy
能量中不能转化为有用功的部分。

174. 压焓图　pressure enthalpy chart

以压力和焓值为坐标，表示物质状态变化的热力状态图。

175. 焓熵图　enthalpy entropy chart
以焓和熵为坐标，表示物质状态变化的热力状态图。

176. 压容图　pressure volume chart
以压力和比容为坐标，表示物质状态变化的热力状态图。

177. 工质　working substance
在热力循环中工作的物质。

178. 制冷剂　refrigerant
制冷系统中，完成制冷循环的工作物质。

179. 冷剂水　water as refrigerant
在吸收式制冷中，作为制冷剂的水。

180. 载冷剂　secondary refrigerant; refrigerating medium
间接制冷系统中，用以吸收被制冷空间或介质的热量，并将其转移给制冷剂的一种流体，也称冷媒。

181. 浓溶液　strong solution; strong liquor
溶质组分较高的溶液。

182. 稀溶液　weak solution
溶质组分较低的溶液。

183. 缓蚀剂　corrosion inhibitor; anticorrosive
加入液体中以降低凝固点的一种添加剂。

184. 闪发气体　flash gas
液态制冷剂由于突然降压而形成的部分蒸发气体。

185. 不凝性气体　non condensable gas; foul gas
在冷凝温度和压力下不凝结而存在于制冷系统中的气体。

186. 热力循环　thermodynamic cycle
工质经过若干热力变化又恢复到初始状态的一系列热力过程。

187. 可逆循环　reversible cycle
由一系列可逆过程所组成的理想热力循环。

188. 卡诺循环　Carnot cycle
包括两个等温过程和两个绝热过程，将热能最大限度地转化为机械能的一种理想的可逆循环。

189. 逆卡诺循环　reverse Carnot cycle
与卡诺循环的过程相同而方向相反的循环。

190. 制冷循环　refrigerating cycle
制冷系统中，制冷剂所经历的一系列热力过程所组成的热力循环。

191. 节流膨胀　throttling expansion
制冷剂通过任何降压元件的膨胀，过程中与外界无机械功的传递。

192. 过冷　subcooling
液态制冷剂的温度降低到相应压力的冷凝温度以下的现象。

193. 过冷度　degree of subcooling
冷凝温度与过冷温度之差。
194. 过热　superheat
气态制冷剂的温度上升到相应压力的饱和温度以上的现象。
195. 过热度　degree of superheat
过热温度与饱和温度之差。
196. 发生器　generator
吸收式制冷机中，通过加热析出制冷剂的设备。
197. 吸收器　absorber
吸收式制冷机中，通过浓溶液吸收剂在其中喷雾以吸收来自蒸发器的制冷剂蒸气的设备。
198. 热泵　heat pump
能实现蒸发器与冷凝器功能转换的制冷机。
199. 计算参数　design conditions
特指设计计算过程中所采用的表征空气状态或变化过程及太阳辐射的物理量，常用的计算参数有干球温度、湿球温度、含湿量、比焓、风速和压力等。
200. 吸收式制冷机　absorption-type refrigerating machine
利用热能完成制冷剂循环和吸收剂循环的制冷机。
201. 压缩式制冷机　compression-type refrigerating machine
用机械压缩制冷剂蒸气完成制冷循环的制冷机。
202. 压缩式冷水机组　compression-type water chiller
将压缩机、冷凝器、蒸发器以及自控元件等组装成一体，可提供冷水的压缩式制冷机。
203. 压缩冷凝机组　condensing unit
将制冷压缩机、冷凝器以及必要的附件等，组装在一个基座上的机组。
204. 活塞式压缩机　reciprocating compressor
靠一个或数个在气缸内作往复运动的活塞，改变其内部容积的压缩机，也称往复式压缩机。
205. 螺杆式压缩机　screw compressor
依靠两个螺旋形转子相互啮合进行压缩的回转式压缩机。
206. 离心式压缩机　centrifugal compressor
利用叶轮旋转产生的离心作用，提升制冷剂气体压力的压缩机。
207. 溴化锂吸收式制冷机　lithium-bromide absorption-type refrigerating machine
以水作制冷剂，以溴化锂作吸收剂完成吸收式制冷循环的制冷机。
208. 室内外计算参数　indoor and outdoor-design conditions
设计计算过程中所采用的室内空气计算参数、室外空气计算参数和太阳辐射照度参数的统称。
209. 干球温度　dry-bulb temperature
干球温度表所指示的温度。

210. 湿球温度　wet-bulb temperature
湿球温度表所指示的温度。

211. 黑球温度　black globe temperature
黑球温度表所指示的温度。

212. 露点温度　dew-point temperature
在大气压力一定、某含湿量下的未饱和空气因冷却达到饱和状态时的温度。

213. 空气湿度　air humidity
表征空气中水蒸气含量多少或干湿度的物理量。

214. 绝对湿度　absolute humidity
单位体积的湿空气中所含水蒸气的质量。

215. 相对湿度　relative humidity
空气实际的水蒸气分压力与同温度下饱和状态空气的水蒸气分压力之比，用百分率表示。

- **燃气工程**

216. 城镇燃气　city gas；town gas
指符合规范的燃气质量要求，供给居民生活、商业、（公共建筑）和工业企业生产作燃料用的功用性质的燃气。城镇燃气一般包括天然气、液化石油气和人工煤气。

217. 城镇燃气工程　city gas engineering
城镇燃气的生产、输配和有关应用的工程。

218. 天然气　natural gas
蕴藏在地层中的可燃性气体。主要是低分子量烷烃类混合物，有些含有 N_2、CO_2、H_2S、H_2 及少量 He 等气体。天然气可分为四种：纯气田天然气、石油伴生气、凝析气田气及矿井气。

219. 纯气田天然气　field natural gas
从纯气田气井中采出的可燃气体，其组成以甲烷为主，还有少量的 N_2、CO_2、H_2S、H_2 或 He 等气体成分。一般不含或少含液相（一般为石油、水）产物。

220. 石油伴生气　associated gas
在石油开采过程中，随着压力的降低，从液相中释放出的可燃气体。其成分多以甲烷为主，还有 C_2、C_3、C_4 及 C_5 等烷烃组分。

221. 凝析气田气　alistillate gas
从气井开采出来经凝析后以甲烷、乙烯为主的可燃气体。

222. 矿井气　mine drainage gas
从井下煤层抽出可燃成分以甲烷为主的可燃气体。甲烷含量随采气方式而变化。

223. 人工燃气　manufactured gas
以固体、液体或气体燃料为原料经转化制得的可燃气体。

224. 煤制气；煤气　coal gas
以煤为原料制得的可燃气体。

225. 油制气　oil gas
以重油、柴油或石脑油等为原料制得的可燃气体。

226. 液化石油气　liquefied petroleum gas ；LPG

在开采和炼制原油过程中，作为副产品而获得的以 C_3、C_4 为主要成分的碳氢化合物。

227. 饱和蒸汽压　saturated vapour pressure

在一定温度下，密闭容器中的液体及其蒸汽处于动态平衡时蒸汽的绝对压力。

228. 沸点　boiling point

液体的饱和蒸汽压等于液体所受压力时的温度，通常指液体的饱和蒸汽压力 101.325kPa 时的温度。

229. 露点　dew point

饱和蒸气经冷却或加压，遇到接触面或凝结核液化成露时的温度。

230. 爆炸极限　explosive limits

可燃气体与空气的混合物遇明火引起爆炸的可燃气体浓度范围。

231. 爆炸上限　upper explosive limit

可燃气体与空气的混合物遇明火引起爆炸的可燃气体最高浓度。

232. 爆炸下限　lower explosive limit

可燃气体与空气的混合物遇明火引起爆炸的可燃气体最低浓度。

- 建筑环境与设备专业常用实验仪器术语

233. 便携式流速仪　flow tracker

234. 皮托管　pitot tube

235. 数字压力计　digital pressure gauge; digital manometer

236. 倾斜式微压计　inclined tube gauge

237. 热球风速仪　hot-bulb anemometer

238. 水静压强仪　water static pressure measuring set

239. 伯努利方程仪　bernoulli equation

240. 沿程阻力系数测定实验台　measure of on-way resistance coefficient laboratory furniture

241. 局部阻力系数测定实验台　measure of coefficient of local resistance laboratory furniture

242. 涡轮流量计　turbine flowmeter

243. 测氧仪　oxymeter

244. 恒温式微机热量计　microcomputer lsothermal point

245. 尘埃粒子计数器　dust particle counter

246. 微电脑激光粉尘仪　microcomputer laser dust monitor

247. 差压检测仪　differential pressure detector

248. 声级计　sound lever meter

249. 数显不锈钢内胆鼓风干燥箱　digital display blast drying oven with stainless inner container

250. 太阳辐射监测站　solar radiation monitoring station

251. 烟尘烟气测定仪　dust and smoke detector

252. 烟雾发生器　smoke generator

253. O_2，CO_2 气体测定仪 O_2，CO_2 gas meter
254. 噪声频谱分析仪   noise spectrum analyzer
255. 立柜式水冷空调机组   upright water-cooled air conditioning unit
256. 管道泵   piping pump
257. 水泵   pater pumps
258. 照度计   illuminance meter
259. 调压器   voltage regulator
260. 污水潜水泵   submersible sewage pump
261. 空气幕   air curtain
262. 自动烟尘测试仪   automatic soot measuring instrument
263. 烟气综合分析仪   flue gas integrated analyzer
264. 智能气体采样器   intelligent gas sampler
265. 汽车排气分析仪   emission analyzer
266. 手持 GPS 定位仪   GPS portable orientation
267. 电热恒温干燥箱   electrothermal constant-temperature dry box
268. 高温箱式电阻炉   high-temperature box resistance furnace
269. 真空干燥箱   vacuum oven
270. 数控超声波清洗器   digital ultrasound cleaner
271. 大气压力表   barometric pressure
272. 热电偶效验仪   thermocouple efficacy instrument
273. 中温法向辐射率测量仪   median temperature normal direction radiance test equipment
274. 气体定压比热测定仪   air specific heat at constant pressure tester
275. 制冷压缩机性能测试仪   refrigerating compressor performance testing
276. 二氧化碳 P-V-T 关系仪 CO_2 P-V-T relations instrument
277. 电动差压变送器   electric-differential pressare transducer
278. 水位仪   limnimeter
279. 万用电源   versatile power
280. 雷诺仪   renault instrument
281. 旋涡仪   swirlmeter
282. 潜水泵   submerged pump
283. 氧弹热量仪   oxygen bomb calorimeter
284. 立式坩锅炉   vertica crucible oven
285. 室外气象测定仪   outdoor weather meter
286. 鼓风干燥箱   air dry oven
287. 恒温水浴   thermostatic waterbath
288. 马福炉   muffle furnace
289. 超级恒温水浴锅   super constant temperature water bath
290. 空气压缩机   air compressor

291. 铂电阻数字温度计　platinum resistor digital thermometer
292. 粉尘分级仪　dust classifier
293. 热球电风速计　hot-bulbanemometer
294. 智能热球电风速计　intelligent hot-bulbanemometer
295. 红外测温仪　infrared thermometer; infrared thermoscope
296. 气溶胶发生器　aerosol dispenser
297. 激光粒子计数器　laser particle counter
298. 大气采样器　air sampler
299. 空气采样泵　air samping pump
300. 电光分析天平　electric-light analytical balance
301. 烘干机　dryer
302. 流量积算仪　flow totalizer
303. 热电偶　thermocouples
304. 热电堆　thermoelectric pile

Appendix B　Introduction to SCI、EI and ISTP（SCI、EI 与 ISTP 检索工具简介）

1. 三大检索基本知识

科技部下属的"中国科学技术信息研究所"从 1987 年起，每年以国外四大检索工具 SCI、ISTP、EI、ISR 为数据源进行学术排行。由于 ISR（《科学评论索引》）收录的论文与 SCI 有较多重复，且收录我国的论文偏少。因此，自 1993 年起，不再把 ISR 作为论文的统计源。而其中的 SCI、ISTP、EI 数据库就是图书情报界常说的国外三大检索工具。

SCI，即《科学引文索引》，是自然科学领域基础理论学科方面的重要的期刊文摘索引数据库。它创建于 1961 年，创始人为美国科学情报研究所所长 Eugene Garfield。利用它，可以检索数学、物理学、化学、天文学、生物学、医学、农业科学以及计算机科学、材料科学等学科方面自 1945 年以来重要的学术成果信息；SCI 还被国内外学术界当作制定学科发展规划和进行学术排名的重要依据。

ISTP，即《科学技术会议录索引》，创刊于 1978 年，由美国科学情报研究所编制，主要收录国际上著名的科技会议文献。它所收录的数据包括农业、环境科学、生物化学、分子生物学、生物技术、医学、工程、计算机科学、化学、物理学等学科。1990~2003 年间，ISTP 和 ISSHP（后文将要讲到 ISSHP）共收录了 60000 个会议的近 300 万篇论文的信息。

EI，即《工程索引》，创刊于 1884 年，由 Elsevier Engineering Information Inc. 编辑出版。主要收录工程技术领域的论文（主要为科技期刊和会议论文），数据覆盖了核技术、生物工程、交通运输、化学和工艺工程、照明和光学技术、农业工程和食品技术、计算机和数据处理、应用物理、电子和通信、控制工程、土木工程、机械工程、材料工程、石油、宇航、汽车工程等学科领域。

2. 三大检索与索引体系

（1）目录和索引的概念及关系

目录是著录一批相关文献（图书、期刊等），并按照一定的次序编排而成的一种揭示与报导文献的工具。目，指文献的篇目名称；录，指文献的内容简介。例如：图书馆的联机公共书目、全国期刊联合目录，OCLC WorldCat、Ulrich's International Periodicals Directory。

索引是将文献（图书、期刊等）中的篇目、语词、主题、人名、地名、事件及其他事物名称，按照一定的方式编排，并指名出处的一种检索工具。目录与索引均属二次文献的范畴，都是用来帮助读者利用一次文献（又称"原始文献"）的；细微的区别在于：目录揭示文献的整体，索引揭示文献的局部。

（2）索引的分类

按照索引的编排方式可以分为：形式索引（著者索引、机构索引、号码索引）、内容索引（分类索引、主题索引、语词索引）和关系索引（引文索引、再版索引）。按照索引的对象（即索引所揭示的原始文献）可以分为：专著索引、报刊索引、会议录索引等。

(3) 三大检索工具在索引体系中的位置

从索引的编排方式来看，SCI 属于关系索引，同时兼具形式索引和内容索引的特征；ISTP 和 EI 具有形式索引和内容索引的特征。

从索引的对象来看，SCI 揭示的是期刊中的论文；ISTP 揭示的是会议中的论文；EI 则兼而有之。

Appendix C   Main Courses for HVAC in Some Famous Universities and Institutes (相关大学 HVAC 课程一览表)

1. 美国普度大学（Purdue University）课程一览表

（1）普度大学本科教育课程一览（Undergraduate-Level Course Sites）
- Thermodynamics
- Introduction to Mechanical Engineering Design
- Basic Mechanics
- Sophomore Seminar (Communications)
- Fluid Mechanics
- Heat and Mass Transfer
- Mechanics of Materials
- Machine Design
- Principles and Practice of Manufacturing Processes
- Systems and Measurements
- System Modeling and Analysis
- Noise Control
- Gas Turbine Engines
- Internal Combustion Engines
- Computer-Aided Design
- Machine Design
- Engineering Design
- Automatic Control Systems

（2）普度大学本科-研究生教育课程一览表（Dual-Level Course Sites）
- Statistical Thermodynamics
- Intermediate Heat Transfer
- Intermediate Fluid Mechanics
- Gas Dynamics
- Engineering Acoustics
- Combustion
- Product and Process Design
- Design for Manufacturability
- Kinematics
- Optimal Design
- Advanced Dynamics
- Mechanical Vibrations

- Vibrations of Discretized Systems
- Vehicle Dynamics
- Mechanical Behavior of Materials
- Machine Design
- Design and Analysis of Robotic Manipulators
- Theory and Design of Control Systems
- Computer Control of Manufacturing Processes
- Digital Control
- Fourier Methods in Digital Signal Processing
- Nonlinear Engineering Systems
- Numerical Methods
- Microprocessors in Electromechanical Systems
- Engineering Optics
- Fundamentals of Noise Control
- Heat Transfer in Electronic Systems
- Kinematics and Dynamics of Human Motion
- Composites and Polymer Processing
- Micromechanics of Materials
- Particle Technology

(3) 普度大学研究生教育课程一览表（Graduate-Level Course Sites）
- Convection of Heat and Mass
- Numerical Methods in Heat, Mass, and Momentum Transfer
- Boundary Layer Theory
- Principles of Turbulence
- Advanced Engineering Acoustics
- Computational Fluid Dynamics
- Aeroacoustics
- Bifurcations and Chaos
- Advanced Engineering Optics
- Intelligent Systems

2. 英国卡迪夫大学（Cardiff University）相关专业教育课程一览表

- Core modules (60 credits)
- Site and Environment
- Earth and Society
- Building Fabric
- Performance Evaluation
- Low Carbon Footprint
- Efficient Services

- Specialist modules (60 credits)
- Building Performance Modelling
- Environmental Design Modelling
- Explorations in modelling
- Dissertation module (60 credits)

3. 英国诺丁汉大学（the University of Nottingham）相关专业（Architecture and Built Environment）教育课程一览表

(1) 诺丁汉大学本科教育主要课程一览（Undergraduate-Level Course Sites）
- Architectural Environment Engineering BEng Hons
- Architecture (BArch Hons/Dip Arch) BArch Hons/Dip Arch
- Architecture (DipArch) DipArch
- Architecture and Environmental Design MEng Hons/Dip Arch
- Sustainable Built Environment BA Hons

(2) 诺丁汉大学研究生教育主要课程一览表（Graduate-Level Course Sites）

Architecture and Urban Design
- MArch in Design
- MArch in Sustainable Tall Buildings
- MArch in Environmental Design
- MArch in Urban Design
- MArch in Technology
- MArch in Theory and Design
- MA in Architecture and Critical Theory
- ARB/RIBA Part 3 Professional Practice

Building Technology and Sustainable Energy Technologies
- MSc in Sustainable Building Technology
- MSc in Renewable Energy and Architecture
- MSc in Sustainable Energy and Entrepreneurship
- MSc in Energy Conversion and Management

4. 加拿大多伦多大学（University of Toronto）相关专业（Mechanical Engineering）教育课程一览表

(1) 多伦多大学本科教育主要课程一览（Undergraduate-Level Course Sites）
- Dynamics
- Introduction to Mechanical and Industrial Engineering
- Essays in Technology and Culture
- Thermodynamics
- Manufacturing Engineering
- Mechanics of Solids

- Engineering Analysis
- Probability and Statistics with Engineering Applications
- Algorithms and Numerical Methods
- Statistics and Design of Experiments
- Human Centred Systems Design
- Data Modelling
- Engineering Economics and Accounting
- Operational Research
- Kinematics and Dynamics of Machines
- Mechanical and Thermal Energy Conversion Processes
- Fluid Mechanics
- Heat and Mass Transfer
- Design for the Environment
- Mechanics of Solids
- Physiological Control Systems
- Engineering Physics
- Computer Aided Design
- Circuits with Applications to Mechanical Engineering Systems
- Industrial Ergonomics and the Workplace
- Ergonomic Design of Information Systems
- Case Studies in Ergonomics
- Analog and Digital Electronics for Mechatronics
- Design and Analysis of Information Systems
- Business Process Engineering
- Organization Design
- Systems Modelling and Simulation
- Resource and Production Modelling
- Methods of Quality Control and Improvement
- Cases in Operational Research
- Ecological Systems
- Vibrations
- Control Systems
- Nuclear Engineering I: Reactor Physics and the Nuclear Fuel Cycle
- Nuclear Engineering II: Thermal and Mechanical Design of Nuclear Power Reactors
- Thermal Energy Conversion
- Applied Fluid Mechanics
- Automated Manufacturing
- Microprocessors and Embedded Microcontrollers
- Biomechanics

- Mechanical Design: Theory and Methodology
- Machine Design
- Mechatronics Systems: Design and Integration
- Mechatronics Principles
- Electromechanical Energy Conversion
- Engineering Psychology and Human Performance
- Human Computer Interface Design for Complex Systems
- Decision Support Systems
- Knowledge Modelling and Management
- Manufacturing and Production Systems
- Integrated System Design
- Smart Materials and Structures
- Facility Planning
- Reliability and Maintainability Engineering
- Entrepreneurship and Business for Engineers
- Thesis
- MEMS Design and Microfabrication
- Problems in Heat Transfer
- Alternative Energy Systems
- Combustion and Fuels
- Fuel Cell Systems
- Fundamentals of Aircraft Design
- Engineering Analysis
- Biomechanics
- Product Design
- Healthcare Systems
- Scheduling
- Decision Analysis

(2) 多伦多大学研究生教育主要课程一览表 (Graduate-Level Course Sites)

Thermal Sciences
- Thermodynamics
- Statistical Thermodynamics
- Non-Equilibrium Thermodynamics
- Conduction Heat Transfer
- Heat Transfer with Phase Change
- Partially Ionized Gases
- Combustion Engine Processes
- Fundamentals of Combustion
- Diffusion Waves

- Engineering applications of sound, electromagnetic, thermal and photonic waves
- Problems in Heat Transfer
- Alternative Energy Systems
- Combustion and Fuels
- Fuel Cell Systems

Fluid Mechanics
- Fluid Mechanics
- Non-Newtonian Fluid Mechanics
- Structure of Turbulent Flows
- Computational Fluid Mechanics and Heat Transfer
- Convective Heat and Mass Transfer
- Environmental Fluid Dynamics
- Multiphase Flows
- Microfluidics and Laboratory-on-a-Chip Systems

Energy Studies
- Materials for Clean Energy

5. 美国加州大学伯克利分校（University of California——Berkeley）相关专业（Mechanical Engineering）课程一览表

（1）加州大学伯克利分校教育课程一览（Undergraduate-Level Course Sites）
- Freshman Seminars
- Thermodynamics
- Introduction to Solid Mechanics
- Supervised Independent Group Studies
- High Mix/Low Volume Manufacturing
- Introduction to Measurement Systems for Mechatronics
- Mechatronics Design
- Engineering Mechanics
- Thermodynamics and Biothermodynamics
- Fluid Mechanics
- Mechanical Engineering Laboratory
- Mechanical Behavior of Engineering Materials
- Heat Transfer
- Introduction to Product Development
- Structural Aspects of Biomaterials
- Introduction to Nanotechnology and Nanoscience
- Introduction to MEMS (Microelectromechanical Systems)
- Processing of Materials in Manufacturing
- Composite Materials—Analysis, Design, Manufacture

- Computer-Aided Mechanical Design
- Design of Planar Machinery
- Vehicle Dynamics and Control
- Dynamic Systems and Feedback
- Mechanical Vibrations
- Automatic Control Systems
- Feedback Control Systems
- Design of Microprocessor-Based Mechanical Systems
- Combustion Processes
- Energy Conversion Principles
- Advanced Heat Transfer
- Engineering Aerodynamics
- Marine Statics and Structures
- Ocean-Environment Mechanics
- Fluid Mechanics of Biological Systems
- Microscale Fluid Mechanics
- Engineering Mechanics
- Fundamentals of Acoustics
- Intermediate Dynamics
- Orthopedic Biomechanics
- Laboratory in the Mechanics of Organisms
- Engineering Analysis Using the Finite Element Method
- Introduction to Continuum Mechanics
- Professional Communication for Mechanical Engineers
- Practical Control System Design: A Systematic Loopshaping Approach
- Model Predictive Control
- Practical Control System Design: A Systematic Optimization Approach
- Honors Undergraduate Research
- Directed Group Studies for Advanced Undergraduates
- Supervised Independent Study

(2) 加州大学伯克利分校研究生教育课程一览表 (Graduate-Level Course Sites)
- Heat and Mass Transport in Biomedical Engineering
- Fluid Mechanics of Biological Systems
- Advanced Tissue Mechanics
- Biomimetic Engineering — Engineering from Biology
- Introduction to MEMS Design
- Parametric and Optimal Design of MEMS
- Precision Manufacturing
- High-Tech Product Design and Rapid Manufacturing

- Advanced Manufacturing Processes
- Polymer Engineering
- Mechanical Behavior of Engineering Materials
- Deformation and Fracture of Engineering Materials
- Tribology
- Mechanical Behavior of Composite Materials
- Computer-Aided, Optimal Mechanical Design
- Design of Basic Electro-Mechanical Devices
- Real-Time Applications of Mini and Micro Computers
- Advanced Control Systems
- Multivariable Control System Design
- Control and Optimization of Distributed Parameters Systems
- Control of Nonlinear Dynamic Systems
- Advanced Design and Automation
- Advanced Marine Structures
- Marine Hydrodynamics
- Advanced Methods in Free-Surface Flows
- Heat Conduction
- Heat Convection
- Thermal Radiation
- Thermodynamics
- Combustion
- Advanced Combustion
- Heat Transfer with Phase Change
- Microscale Thermophysics and Heat Transfer
- Advanced Fluid Mechanics
- Theory of Fluid Sheets and Fluid Jets
- Turbulence
- Dynamics and Stability of Engineering and Geophysical Flows with Rotation, Convection, or Waves
- Physicochemical Hydrodynamics
- Oscillations in Linear Systems
- Random Oscillations of Mechanical Systems
- Advanced Dynamics
- Oscillations in Nonlinear Systems
- Statistical Mechanics of Elasticity
- Introduction to the Finite Element Method
- Finite Element Methods in Nonlinear Continua
- Methods of Tensor Calculus and Differential Geometry

- Theory of Elasticity
- Wave Propagation in Elastic Media
- Nonlinear Theory of Elasticity
- Foundations of the Theory of Continuous Media
- Surfaces of Discontinuity and Inhomogeneities in Deformable Continua
- Electrodynamics of Continuous Media
- Theory of Plasticity
- Multiscale Modeling and Design of New Materials
- Theory of Elastic Stability
- Theory of Shells
- Nonlinear Dynamics of Continuous Systems
- Topics in Fluid Mechanics
- Solid Modeling
- Laser Processing and Diagnostics
- Green Product Development: Design for Sustainability
- Predictive Control for Linear and Hybrid Systems
- Introduction to Nano-Biology
- Expert Systems in Mechanical Engineering
- System Identification
- New Product Development: Design Theory and Methods
- Dynamic Control of Robotic Manipulators
- Topics in Manufacturing
- Hybrid Systems and Intelligent Control
- Plasmonic Materials
- Advanced Technical Communication: Proposals, Patents, and Presentations
- Topics in Control, Modeling and Optimization
- Group Studies, Seminars, or Group Research
- Individual Study or Research
- Individual Study for Doctoral Students

6. 宾夕法尼亚大学 (University of Pennsylvania) 相关专业 (Mechanical Engineering and Applied Mechanics) 教育课程一览表

- Independent Study
- Introduction to Mechanical Design
- Introduction to Mechanics
- Visual Thinking
- Introduction to Mechanics Lab
- Fundamentals of Mechanical Protoyping
- Thermodynamics

- Statics and Strength of Materials
- Engineering Mechanics: Dynamics
- Elements of Mechanical Engineering Design
- Introduction to Flight
- Mechanical Engineering Laboratory
- Energy Systems, Resources and Technology
- Fluid Mechanics
- Design of Thermal/Fluid Systems
- Vibrations of Mechanical Systems
- Heat and Mass Transfer
- Mechanical Engineering Design Laboratory
- Mechanics of Solids
- Energy Engineering
- Mechanical Properties of Macro/Nanoscale Materials
- Design of Mechatronic Systems
- Robotics
- Mechanical Engineering Design Projects
- Mechanics of Materials
- Continuum Biomechanics
- Mechanics of Human Motion
- Creative Thinking and Design
- Modern Feedback Control Theory
- Design for Manufacturability
- Product Design
- Advanced Mechatronic Reactive Spaces
- Elasticity and Micromechanics of Materials
- Robotics and Automation
- Introduction to Parallel Computing
- Fundamentals of Sensor Technology
- Finite Element Analysis
- Advanced Kinematics
- Introduction to MEMS and NEMS
- Continuum Mechanics
- Advanced Heat and Mass Transfer
- Advanced Dynamics
- Viscous Fluid Flow
- Nanomechanics and Nanotribology at Interfaces
- Optimal Design of Mechanical Systems
- Continuum Biomechanics

- Aerodynamics
- Micro-Electro-Mechanical Systems
- Nanoscale Systems Biology
- The Principles and Practice of Microfabrication Technology
- Transport Processes
- Advanced Topics in Transport Phenomena
- Micro/Nanoscale Energy Transport
- Physicochemical Hydrodynamics and Interfacial Phenomena
- Advanced Mechatronics
- Nonlinear Control Theory
- Robotics
- Haptic Interfaces
- Advanced Continuum Mechanics
- Advanced Elasticity
- Plasticity
- Fracture Mechanics
- Rods and Shells
- Composite Materials
- BioTransport: Fluid Mechanics, Heat and Mass Transfer
- Computational Mechanics
- Fundamentals of Complex Fluids
- Atomistic Modeling in Materials Science
- Advanced Thermodynamics Seminar
- Advanced Molecular Thermodynamics
- Entropic Forces in Biomechanics
- Heat Conduction and Mass Diffusion
- Heat Transfer II: Convection
- Heat Transfer III: Radiation
- Advanced topics in solid mechanics, dynamics, thermal-fluid science, or energy disciplines
- Special Topics in Mechanics of Materials
- Topics in Mechanical Systems
- SM 699. MEAM Seminar
- Teaching Practicum
- Masters Thesis
- Dissertation
- Thesis/Dissertation Research

Appendix D Key Chinese Laws, Regulations and Standards Pertaining to Urban Housing Development in Xi'an (HVAC 相关中国法律、规范及标准)

Areas	Laws	Regulations/Standards	Local Regulations/Standards
Planning	Law of the P. R. China on - Protection of Cultural Relics (1982) Protection of Environment (1998) Urban Planning (1989) Urban Real Estates Development (1995) Environment Impact Assessment (2002) Land Resource Management (1986&2004) Property Ownership (2007)	Enacted by the State Council: Regulations- Urban Plantation (1992) Urban Water Supply (1994) Management of Urban Real Estate Development (1998) Management of Construction Project Surveying and Design (2000) Management of Scenic and Heritage Sites (2006) Enacted by the Ministry of Construction: Rule or Measure on- Urban Water-saving Management (1989) Compilation and Approval of Town and City Planning System, 1994, and its Implementation Guideline (1995, Complication of Urban Planning, 2005) Management of Urban Natural Gas Supply (1997) Management of Urban Planning for Earthquake Preparedness and disaster Prevention (2003) Quality Control of Construction Project Surveying (2003) Management of Urban Public Buses and Coach Transportation (2005) Earthquake Preparedness of Building Construction (2006) Management of Urban Green Structure (2002) Management of Urban Purple Structure (2003) Management of Urban Blue Structure (2006) Management of Urban Yellow Structure (2006) Domestic Building Energy Efficiency (2006) Enacted by other central government bodies: National Standard on Urban Planning Terminology (1992)	Enacted by the Shanxi provincial government: Measure or Decree of Shanxi Province on- Implementation of Urban Planning Law (1991) Implementation of Urban Plantation Regulation (1995) Implementation of Land Resource Management Law (2000) Land Requisition for Construction Projects (2001) Management of Urban Real Estate Market (1995&2004) Implementation Guideline of Housing Scheme for Low-income family (2005) Plan Compilation of Urban Housing Development (2006) Protection of Cultural Relics (2006) Building Energy Efficiency (2007) Enacted by the Xi'an city government: Measure or Decree of Xi'an City on- Management of Urban Planning (1993) Management of Urban Real Estate Market (2004) Protection of Historic and Construction Data (2005&2006) Management of Urban Planning (2005) Management of Geological Enviroment (2005)

continued

Areas	Laws	Regulations/Standards	Local Regulations/Standards
Design	Law of the P. R. China on - Protection of Environment (1988) Construction (1996) Urban Planning (1989) Project Tendering and Bidding (1999) Contract (1999) Road and Transportation Safety (Parking facilities, 2003)	Enacted by the State Council: Regulations- Management of Construction Project Surveying and Design (2000) Quality Management of Construction Project (2000) Enacted by the Ministry of Construction: Rule or Measure on- Tendering and Bidding Management of Construction Project (2000) Management of Construction and Design Plan & Documentation of Residential Building and Municipal Infrastructure (2006) Management of Urban Yellow Structure (2006) Domestic Building Energy Efficiency (2006) Design Guidelines on or Code for- Thermal Performance of Domestic Building (1993) Domestic Building Energy Efficiency (Heating, 1996) Electricity Supply (1995) Residential Building (1999) Residential Building Energy Efficiency in the Hot Summer and Cold Winter Regions (2001) Building Natural Lighting (2001) Heating, Ventilation and Air Conditioning (2003) Building Illuminating (2004) Application of Solar Heating Water System in Domestic Building (2005) Building Structure Load (2002, 2006) Building Fire Safety (2006) Indoor Air Pollution Control in Domestic Building (2001, 2006) Masonary Filler Wall Structure (2006) Enacted by other central government bodies: Design Guidelines on- Domestic Buildings (1987) Soundproof in Domestic Buildings (1995, 2001)	Enacted by the Shaanxi provincial government: Measure or Decree of Shaanxi Province on- Management of Monitoring Fire Safety of Construction Project (1998) Implementation of Tendering and Bidding Law (2004) Management of Urban Real Estate Market (2004) Key Areas in Reviewing the Construction Plan regarding Building Energy Efficiency (2004) Implementation Guideline on the "Design Guideline on Domestic Building Energy Efficiency" (1997) Temporary Guideline on House-hold Based Heating Control and metering (2005) Enacted by the Shaanxi provincial government: Xi'an City Management Guideline on Construction Project Surveying and Design (2005) Xi'an City Design Standard on Residential Building Energy Efficiency (2007)

continued

Areas	Laws	Regulations/Standards	Local Regulations/Standards
Construction	Law of the P. R. China on - Protection of Environment (1988) Labour (1991) Construction (1996) Project Tendering and Bidding (1999) Contract (1999) Air Pollution Prevention (2000) Safety in Working Environment (2002) Road and Transportation Safety (Road digging and working (2003))	Enacted by the State Council: Regulations- Quality Management of Construction Project (2000) Site Safety Management of Construction Project (2003) Safety Licence in Working Places (2004) Enacted by the Ministry of Construction: Rule or Measure on- Management of Comprehensive Inspection and Commissioning of New Urban Residential Development (1993) Management of Building Construction Licence (1999) Management of Construction Tendering and Bidding in Residential Building and Municipal Infrastructure Projects (2001) Management of Construction Sub-contracting in Residential Building and Municipal Infrastructure Projects (2004) Working Safety Licence Management of Construction Company (2004) Construction Project Safety Management, Well-manned Site Construction Fee and its Management (2005) Urban Construction Waste Management (2005) Construction Project Quality Inspection (2005) Management of Urban Sewage Licence (2006) Domestic Building Energy Efficiency (2006) Construction Project Management (2006) Code for the Assessment of Building Project Construction Quality (2006) Guideline on Green Construction (2007)	Enacted by the Shaanxi provincial government: Measure or Decree of Shaanxi Province on- Implementation of Urban Plantation Regulation (1995) Indoor Decoration Management (2002) Implementation of Tendering and Bidding Law (2004) Management of Urban Real Estate Market (1995&2004) Enacted by the Xi'an city government: Measure or Decree of Xi'an City on- Construction Waste Management (2003), and its Implementation Guideline (2003) Indoor Decoration Management (2006)
Occupancy	Law of the P. R. China on - Property Ownership (2007)	Enacted by the State Council: Property Management Regulation (2003) Enacted by the Ministry of Construction: Rule or Measure on-	Enacted by the Shaanxi provincial government: Measure or Decree of Shaanxi Province on- Implementation of Urban Plantation Regulation (1995) Property Management Fee in Urban Residential Area (temporary, 1998)

continued

Areas	Laws	Regulations/Standards	Local Regulations/Standards
		Inspection of Toilet Tank Application in Urban Domestic Building (1992); Urban Domestic Waste Management (2005); Management of New Urban Residential Area (1994); Management of White Ants Prevention in Urban Building (1999, 2004); Indoor Decoration Management (2002)	Urban Residential Area Property Management (2004); Enacted by the Xi'an city government: Measure or Decree of Xi'an City on- Occupancy Safety Management of Urban Residential Buildings (2004); Urban Residential Area Property Management (2004); Indoor Decoration Management (2006)
Energy	Law of the P. R. China on - Energy Saving (1997); Renewable Energy (2005)	Enacted by the Ministry of Construction: Technical Guideline on the Application of Solar Heating Water System in Domestic Building (2005)	
Water	Law of the P. R. China on - Water (2002)	Enacted by the State Council: Urban Water Supply Measure (1994); Enacted by the Ministry of Construction: Rule or Measure on- Urban Water-saving Management (1989); Inspection of Toilet Tank Application in Urban Domestic Building (1992); Management of Urban Sewage Licence (2006)	
Waste Management (including both construction and domestic waste)	Law of the P. R. China on - Environment Protection from solid Waste Pollution (1995&2005)	Enacted by the Ministry of Construction: Rule or Measure on- Urban Domestic Waste Management (1993); Urban Construction Waste Management (2005); Enacted by the Ministry of Construction:	Enacted by the Xi'an city government: Measure or Decree of Xi'an City on- Urban Domestic Waste Management (2003); Urban Construction Waste Management (2003)

continued

Areas	Laws	Regulations/Standards	Local Regulations/Standards
Air Quality/Indoor Environmental Quality	Law of the P. R. China on Air Pollution Prevention	Regulation on Indoor Environmental Pollution Control in Domestic Building Project (2001&2006)	
Urban Infrastructure (including green/blue structure, transport etc.)		Enacted by the Ministry of Construction: Rule or Measure on— Management of Urban Green Structure (2002) Management of Urban Public Bus and Coach Transportation (2005) Management of Urban Blue Structure (2006) Management of Urban Yellow Structure (2006) Design Guideline on Accessibility of Urban Road and Building (2001)	Enacted by the Shaanxi provincial government: Decree of Shaanxi province on Implementation of Urban Plantation Regulation (1995)
Historic and Cultural Heritage		Enacted by the Ministry of Construction: Management Measure of Urban Purple Structure (2003)	Enacted by the Xi'an city government: Decree of Xi'an City on— Protection of Historic and Cultural City (2005) Management of Geological Enviroment (2005)
Qualification		Enacted by the State Council: Regulation on the Management of Chinese Chartered Architects (1995) Enacted by the Ministry of Construction: Management Measure on— Qualification of Property Management Company (2004) Chartered Surveyor (2005) Construction Project Quality Inspection (2005) Chartered Architect (2006)	
Building Assessement		Chartered Site Inspector (2006) Enacted by the Ministry of Construction: Code for Economic Assessment of Residential Building Technology (1998) Code for the Assessment of Green Buildings (2005)	

参考文献

[1] Stephen Wolfram，*A New Kind of Science*（Wolfram Media，2002）
[2] Peter Salamon，VARIOUS APPROACHES TO THERMODYNAMICS
[3] http：//www. fluent. com/about/cfdhistory. htm
[4] http：//www. pexheat. com/history-radiant-panel-heating
[5] http：//www. bchydro. com/
[6] http：//en. wikipedia. org/wiki/
[7] http：//www. design. asu. edu/radiant/01_thermalComfort/thermalC_main. htm
[8] Thibaut Vitte，Monika Woloszyn，Jean Brau，SOLAR DESSICANT COOLING
[9] EXAMPLE OF APPLICATION TO A LOW ENERGY BUILDING，ANNEX 41 spring meeting，Montréal，16-18 May 2005.
[10] Passive Solar Building Design Guidelines and Recognition Program Prepared by City of Santa Barbara Community Development Department，December 2006.
[11] Photovoltaics：Basic Design Principles and Components，DOE/GO-10097-377，FS 231 March 1997
[12] Olle Törnblom "Introduction course in particle image velocimetry"，March 31，2004
[13] ASHRAE Standard 62. 2，Ventilation and Acceptable Indoor Air Quality in Low-Rise Residential Buildings.
[14] ASHRAE Handbook（2001）—Fundamentals，Chapter 26，Ventilation and Infiltration.
[15] Emmanuelle Gallo，"Skyscrapers and District Heating，an inter-related History 1876-1933."
[16] Refrigeration and air-conditioning [R]. Ashrae Handbook，2009
[17] Ventilation [R]. ASHRAE Handbook，2009
[18] HVAC Fundamentals. Vol. 1：Heating Systems，Furnaces，and Boilers [M]. Audel，2004
[19] Refrigeration Principles And Systems [M]. John Wiley and Sons，1984
[20] Abdeen Mustafa Omer. Ground-source heat pumps systems and applications. Renewable and Sustainable Energy Reviews．12（2008）：344－371
[21] 俞炳丰. 科技英语论文实用写作指南 [M]. 西安：西安交通大学出版社，2003
[22] 张寅平，潘毅群，王馨. 高等学校专业英语系列教材——建筑环境与设备工程专业 [M]. 北京：中国建筑工业出版社，2004
[23] 陆亚俊. 暖通空调 [M]. 北京：中国建筑工业出版社，2002
[24] 丁崇功，寇广孝. 工业锅炉设备 [M]. 北京：机械工业出版社，2005
[25] 吴业正. 制冷原理及设备 [M]. 西安：西安交通大学出版社，1997
[26] 赵三元等. 建筑类专业英语暖通与燃气（第一册）. 北京：中国建筑工业出版社，1997
[27] 向阳. 建筑类专业英语暖通与燃气（第二册）. 北京：中国建筑工业出版社，1997
[28] 张国强. 建筑环境与设备工程专业英语. 长沙：湖南大学出版社，2003
[29] http：//www. efunda. com/formulae/fluids/overview. cfm
[30] http：//www. efunda. com/formulae/heat_transfer/home/overview. cfm
[31] http：//www. britannica. com/EBchecked/topic/258832/heating#
[32] http：//en. wikipedia. org/wiki/Heat_exchanger
[33] http：//en. wikipedia. org/wiki/Central_heating

[34] http://en.wikipedia.org/wiki/District_heating
[35] http://en.wikipedia.org/wiki/Fuel_gas
[36] http://en.wikipedia.org/wiki/Natural_gas